Hot Money

Hot Money

NAOMI KLEIN

PENGUIN BOOKS — GREEN IDEAS

PENGUIN BOOKS

UK | USA | Canada | Ireland | Australia
India | New Zealand | South Africa

Penguin Books is part of the Penguin Random House group of companies
whose addresses can be found at global.penguinrandomhouse.com.

Penguin
Random House
UK

First published in *This Changes Everything* the United States of America
by Simon & Schuster 2014
First published in *This Changes Everything* in Great Britain by Allen Lane 2014
This extract published in Penguin Books 2021

003

Text copyright © Naomi Klein, 2014

Set in 11.5/14pt Dante MT Std
Typeset by Jouve (UK), Milton Keynes
Printed and bound in Great Britain by Clays Ltd, Elcograf S.p.A.

The authorized representative in the EEA is Penguin Random House Ireland,
Morrison Chambers, 32 Nassau Street, Dublin D02 YH68

A CIP catalogue record for this book is available from the British Library

ISBN: 978-0-141-99688-2

www.greenpenguin.co.uk

MIX
Paper | Supporting
responsible forestry
FSC® C018179
www.fsc.org

Penguin Random House is committed to a
sustainable future for our business, our readers
and our planet. This book is made from Forest
Stewardship Council® certified paper.

Contents

Hot Money

How Free Market Fundamentalism
Helped Overheat the Planet

If the climate movement had a birthday, a moment when the issue pierced the public consciousness and could no longer be ignored, it would have to be June 23, 1988. Global warming had been on the political and scientific radar long before that, however. The basic insights central to our current understanding date back to the beginning of the second half of the nineteenth century, and the first scientific breakthroughs demonstrating that burning carbon could be warming the planet were made in the late 1950s. In 1965, the concept was so widely accepted among specialists that U.S. president Lyndon B. Johnson was given a report from his Science Advisory Committee warning that, 'Through his worldwide industrial civilization, Man is unwittingly conducting a vast geophysical experiment . . . The climatic changes that may be produced by the increased CO_2 content could be deleterious from the point of view of human beings.'[1]

But it wasn't until James Hansen, then director of

NASA's Goddard Institute for Space Studies, testified before a packed congressional hearing on June 23, 1988, that global warming became the stuff of chat shows and political speeches. With temperatures in Washington, D.C., a sweltering 98 degrees Fahrenheit (still a record for that day), and the building's air conditioning on the fritz, Hansen told a room filled with sweaty lawmakers that he had '99 percent confidence' in 'a real warming trend' linked to human activity. In a comment to *The New York Times* he added that it was 'time to stop waffling' about the science. Later that same month, hundreds of scientists and policymakers held the historic World Conference on the Changing Atmosphere in Toronto where the first emission reductions were discussed. The United Nations' Intergovernmental Panel on Climate Change (IPCC), the premier scientific body advising governments on the climate threat, held its first session that November. By the following year, 79 percent of Americans had heard of the greenhouse effect – a leap from just 38 percent in 1981.[2]

The issue was so prominent that when the editors of *Time* magazine announced their 1988 'Man of the Year,' they went for an unconventional choice: 'Planet of the Year: Endangered Earth,' read the magazine's cover line, over an image of the globe held together with twine, the sun setting ominously in the background. 'No single individual, no event, no movement captured imaginations or dominated headlines more,' journalist Thomas Sancton explained, 'than the clump of rock and soil and water and air that is our common home.'[3]

More striking than the image was Sancton's accompanying essay. 'This year the earth spoke, like God warning Noah of the deluge. Its message was loud and clear, and suddenly people began to listen, to ponder what portents the message held.' That message was so profound, so fundamental, he argued, that it called into question the founding myths of modern Western culture. Here it is worth quoting Sancton at length as he described the roots of the crisis:

In many pagan societies, the earth was seen as a mother, a fertile giver of life. Nature – the soil, forest, sea – was endowed with divinity, and mortals were subordinate to it. The Judeo-Christian tradition introduced a radically different concept. The earth was the creation of a monotheistic God, who, after shaping it, ordered its inhabitants, in the words of Genesis: 'Be fruitful and multiply, and replenish the earth and subdue it: and have dominion over the fish of the sea and over the fowl of the air and over every living thing that moveth upon the earth.' The idea of dominion could be interpreted as an invitation to use nature as a convenience.[4]

The diagnosis wasn't original – indeed it was a synthesis of the founding principles of ecological thought. But to read these words in America's most studiously centrist magazine was nothing short of remarkable. For this reason and others, the start of 1989 felt to many in the environmental movement like a momentous juncture, as if the thawing of the Cold War and the warming of the planet were together helping to birth a new consciousness, one in which cooperation would triumph over domination, and humility before nature's complexity would challenge technological hubris.

As governments came together to debate responses to climate change, strong voices from developing countries spoke up, insisting that the core of the problem was the high-consumption lifestyle that dominated in the West. In a speech in 1989, for instance, India's President R. Venkataraman argued that the global environmental crisis was the result of developed countries' 'excessive consumption of all materials and through large-scale industrialization intended to support their styles of life.'[5] If wealthy countries consumed less, then everyone would be safer.

But if that was the way 1989 began, it would end very differently. In the months that followed, popular uprisings would spread across the Soviet-controlled Eastern Bloc, from Poland to Hungary and finally to East Germany where, in November 1989, the Berlin Wall collapsed. Under the banner 'the End of History,' right-wing ideologues in Washington seized on this moment of global flux to crush all political competition, whether socialism, Keynesianism, or deep ecology. They waged a frontal attack on political experimentation, on the idea that there might be viable ways of organizing societies other than deregulated capitalism.

Within a decade, all that would be left standing would be their own extreme, pro-corporate ideology. Not only would the Western consumer lifestyle survive intact, it would grow significantly more lavish, with U.S. credit card debt per household increasing fourfold between 1980 and 2010.[6] Simultaneously, that voracious lifestyle would be exported to the middle and upper classes in every corner of the globe – including, despite earlier protestations, India, where it would wreak environmental damage on a scale difficult to fathom. The victories in the new era would be faster and bigger than almost anyone predicted; and the armies of losers would be left to pick through the ever-growing mountains of methane-spewing waste.

Trade and Climate: Two Solitudes

Throughout this period of rapid change, the climate and trade negotiations closely paralleled one another, each winning landmark agreements within a couple of years of each other. In 1992, governments met for the first United Nations Earth Summit in Rio, where they signed the United Nations Framework Convention on Climate Change (UNFCCC), the document that formed the basis for all future climate negotiations. That same year, the North American Free Trade Agreement was signed, going into effect two years later. Also in 1994, negotiations establishing the World Trade Organization concluded, and the new global trade body made its debut the next year. In 1997, the Kyoto Protocol was adopted, containing the first binding emission reduction targets. In 2001, China gained full membership in the WTO, the culmination of a trade and investment liberalization process that had begun decades earlier.

What is most remarkable about these parallel processes – trade on the one hand, climate on the other – is the extent to which they functioned as two solitudes. Indeed, each seemed to actively pretend that the other did not exist, ignoring the most glaring questions about how one would impact the other. Like, for example: How would the vastly increased distances that basic goods would now travel – by carbon-spewing container ships and jumbo jets, as well as diesel trucks – impact the carbon emissions that the climate negotiations were aiming to reduce? How would the aggressive protections for technology patents enshrined under the WTO impact the demands being made by developing nations in the climate negotiations for free transfers of green technologies to help them develop on a low-carbon path? And perhaps most critically, how would provisions that allowed private companies to sue national governments over laws that impinged on their profits dissuade governments from adopting tough anti-pollution regulations, for fear of getting sued?

These questions were not debated by government negotiators, nor was any attempt made to resolve their obvious contradictions. Not that there was ever any question about which side would win should any of the competing pledges to cut emissions and knock down commercial barriers ever come into direct conflict: the commitments made in the climate negotiations all effectively functioned on the honor system, with a weak and unthreatening mechanism to penalize countries that failed to keep their promises. The commitments made under trade agreements, however, were enforced by a dispute settlement system with real teeth, and failure to comply would land governments in trade court, often facing harsh penalties.

In fact, the hierarchy was so clear that the climate negotiators formally declared their subservience to the trading system from the start. When the U.N. climate agreement was signed at the Rio Earth Summit in 1992, it made clear that 'measures taken to combat climate change, including unilateral ones, should not constitute . . . a disguised restriction on international trade.' (Similar language appears in the Kyoto Protocol.) As Australian political scientist Robyn Eckersley puts it, this was 'the pivotal moment that set the shape of the relationship between the climate and trade regimes' because, 'Rather than push for the recalibration of the international trade rules to conform with the requirements of climate protection . . . the Parties to the climate regime have ensured that liberalized trade and an expanding global economy have been protected against trade-restrictive climate policies.' This practically guaranteed that the negotiating process would be unable to reckon with the kinds of bold but 'trade-restrictive' policy options that could have been coordinated internationally – from buy-local renewable energy programs to restrictions on trade in goods produced with particularly high carbon footprints.[7]

A few isolated voices were well aware that the modest gains being made in the negotiations over 'sustainable development' were being actively unmade by the new trade and investment architecture. One of those voices belonged to Martin Khor, then director of the Third World Network, which has been a key advisor to developing country governments in both trade and climate talks. At the end of the 1992 Rio Earth Summit, Khor cautioned that there was a 'general feeling among Southern country delegates . . . that events outside the [summit] process were threatening to weaken the South further and to endanger whatever positive elements exist in' the Rio agenda. The examples he cited were the austerity policies being pushed at the time by the World Bank and the International Monetary Fund, as well as the trade negotiations that would soon result in the creation of the WTO.[8]

Another early warning was sounded by Steven Shrybman, who observed a decade and a half ago that the global export of industrial agriculture had already dealt a devastating blow to any possible progress on emissions. In a paper published in 2000, Shrybman argued that 'the globalization of agricultural systems over recent decades is likely to have been one of the most important causes of overall increases in greenhouse gas emissions.'[9]

This had far less to do with current debates about the 'food miles' associated with imported versus local produce than with the way in which the trade system, by granting companies like Monsanto and Cargill their regulatory wish list – from unfettered market access to aggressive patent protection to the maintenance of their rich subsidies – has helped to entrench and expand the energy-intensive, higher-emissions model of industrial agriculture around the world. This, in turn, is a major explanation for why the global food system now accounts for between 19 and 29 percent of world greenhouse gas emissions. 'Trade policy and rules actually drive climate change in a very structural way in respect of food systems,' Shrybman stressed in an interview.[10]

The habit of willfully erasing the climate crisis from trade agreements continued: for instance, in early 2014, several negotiating documents for the proposed Trans-Pacific Partnership, a controversial new NAFTA-style trade deal spanning twelve countries, were released to the public via WikiLeaks and the Peruvian human rights group RedGE. A draft of the environment chapter had contained language stating that countries 'acknowledge climate change as a global concern that requires collective action and recognize the importance of implementation of their respective commitments under the United Nations Framework Convention on Climate Change (UNFCCC).' The language was vague and nonbinding, but at least it was a tool that governments could use to defend themselves should their climate policies be challenged in a trade tribunal, as Ontario's plan was. But a later document showed that U.S. negotiators had proposed an edit: take out all the stuff about climate change and UNFCCC commitments. In other words, while trade has repeatedly been allowed to trump climate, under no circumstances would climate be permitted to trump trade.[11]

Nor was it only the trade negotiators who blocked out the climate crisis as they negotiated agreements that would send emissions soaring and make many solutions to this problem illegal. The climate negotiations exhibited their own special form of denial. In the early and mid-1990s, while the first climate protocol was being drafted, these negotiators, along with the Intergovernmental Panel on Climate Change, hashed out the details of precisely how countries should measure and monitor how much carbon they were emitting – a necessary process since governments were on the verge of pledging their first round of emission reductions, which would need to be reported and monitored.

The emissions accounting system on which they settled was an odd relic of the pre-free-trade era that took absolutely no account of the revolutionary changes unfolding right under their noses regarding how (and where) the world's goods were being manufactured. For instance, emissions from the transportation of goods across borders – all those container ships, whose traffic has increased by nearly 400 percent over the last twenty years – are not formally attributed to any nation-state and therefore no one country is responsible for reducing their polluting impact. (And there remains little momentum at the

U.N. for changing that, despite the reality that shipping emissions are set to double or even triple by 2050.)[12]

And fatefully, countries are responsible only for the pollution they create inside their own borders – not for the pollution produced in the manufacturing of goods that are shipped to their shores; those are attributed to the countries where the goods were produced.[13] This means that the emissions that went into producing, say, the television in my living room, appear nowhere on Canada's emissions ledger, but rather are attributed entirely to China's ledger, because that is where the set was made. And the international emissions from the container ship that carried my TV across the ocean (and then sailed back again) aren't entered into anyone's account book.

This deeply flawed system has created a vastly distorted picture of the drivers of global emissions. It has allowed rapidly de-industrializing wealthy states to claim that their emissions have stabilized or even gone down when, in fact, the emissions embedded in their consumption have soared during the free trade era. For instance, in 2011, the *Proceedings of the National Academy of Sciences* published a study of the emissions from industrialized countries that signed the Kyoto Protocol. It found that while their emissions had stopped growing, that was partly because international trade had allowed these countries to move their dirty production overseas. The researchers concluded that the rise in emissions from goods produced in developing countries but consumed in industrialized ones was *six times* greater than the emissions savings of industrialized countries.[14]

Cheap Labor, Dirty Energy: A Package Deal

As the free trade system was put in place and producing offshore became the rule, emissions did more than move – they multiplied. Before the neoliberal era, emissions growth had been slowing, from 4.5 percent annual increases in the 1960s to about 1 percent a year in the 1990s. But the new millennium was a watershed: between 2000 and 2008, the growth rate reached 3.4 percent a year, shooting past the highest IPCC projections of the day. In 2009, it dipped due to the financial crisis, but made up for lost time with the historic 5.9 percent increase in 2010 that left climate watchers reeling. (In mid-2014, two decades after the creation of the WTO, the IPCC finally acknowledged the reality of globalization and noted in its Fifth Assessment Report, 'A growing share of total anthropogenic CO_2 emissions is released in the manufacture of products that are traded across international borders.')[15]

The reason for what Andreas Malm – a Swedish expert on the history of coal – describes as 'the early 21st Century emissions explosion' is straightforward enough. When China became the 'workshop of the world' it also became the coal-spewing 'chimney of the world.' By 2007, China was responsible for two thirds of the annual increase in global emissions. Some of that was the result of China's own internal development – bringing electricity to rural areas, and building roads. But a lot of it was directly tied to foreign trade: according to one study, between 2002 and 2008, 48 percent of China's total emissions was related to producing goods for export.[16]

'One of the reasons why we're in the climate crisis is because of this model of globalization,' says Margrete Strand Rangnes, executive vice president at Public Citizen, a Washington-based policy institute that has been at the forefront of the fight against free trade. And that, she says, is a problem that requires 'a pretty fundamental re-formation of our economy, if we're going to do this right.'[17]

International trade deals were only one of the reasons that governments embraced this particular model of fast-and-dirty, export-led development, and every country had its own peculiarities. In many cases (though not China's), the conditions attached to loans from the International Monetary Fund and World Bank were a major factor, so was the economic orthodoxy imparted to elite students at schools like Harvard and the University of Chicago. All of these and other factors played a role in shaping what was (never ironically) referred to as the Washington Consensus. Underneath it all is the constant drive for endless economic growth, a drive that goes much deeper than the trade history of the past few decades. But there is no question that the trade architecture and the economic ideology embedded within it played a central role in sending emissions into hyperdrive.

That's because one of the primary driving forces of the particular trade system designed in the 1980s and 1990s was always to allow multinationals the freedom to scour the globe in search of the cheapest and most exploitable labor force. It was a journey that passed through Mexico and Central America's sweatshop maquiladoras and had a long stopover in South Korea. But by the end of the 1990s, virtually all roads led to China, a country where wages were extraordinarily low, trade unions were brutally suppressed, and the state was willing to spend seemingly limitless funds on massive infrastructure projects – modern ports, sprawling highway systems, endless numbers of coal-fired power plants, massive dams – all to ensure that the lights stayed on in the factories and the goods made it from the assembly lines onto the container ships on time. A free trader's dream, in other words – and a climate nightmare.

A nightmare because there is a close correlation between low wages and high emissions, or as Malm puts it, 'a causal link between the quest for cheap and disciplined labor power and rising CO_2 emissions.' And why wouldn't there be? The same logic that is willing to work laborers to the bone for pennies a day will burn mountains of dirty coal while spending next to nothing on pollution controls because it's the cheapest way to produce. So when the factories moved to China, they also got markedly dirtier. As Malm points out, Chinese coal use was declining slightly between 1995 and 2000, only for the explosion in manufacturing to send it soaring once again. It's not that the companies moving their production to China wanted to drive up emissions: they were after the cheap labor, but exploited workers and an exploited planet are, it turns out, a package deal. A destabilized climate is the cost of deregulated, global capitalism, its unintended, yet unavoidable consequence.[18]

This connection between pollution and labor exploitation has been true since the earliest days of the Industrial Revolution. But in the past, when workers organized to demand better wages, and when city dwellers organized to demand cleaner air, the companies were pretty much forced to improve both working and environmental standards. That changed with the advent of free trade: thanks to the removal of virtually all barriers to capital flows, corporations could pick up and leave every time labor costs started rising. That's why many large manufacturers left South Korea for China in the late 1990s, and it's why many are now leaving China, where wages are climbing, for Bangladesh, where they are significantly lower. So while our clothes, electronics, and furniture may be made in China, the economic model was primarily made in the U.S.A.

And yet when the subject of climate change comes up in discussion in wealthy, industrialized countries, the instant response, very often, is that it's all China's fault (and India's fault and Brazil's fault and so on). Why bother cutting our own emissions when everyone knows that the fast developing economies are the real problem, opening more coal plants every month than we could ever close?[19] This argument is made as if we in the West are mere spectators to this reckless and dirty model of economic growth. As if it was not our governments and our multinationals that pushed a model of export-led development that made all of this possible. It is said as if it were not our own corporations who, with single-minded determination (and with full participation from China's autocratic rulers), turned the Pearl River Delta into their carbon-spewing special economic zone, with the goods going straight onto container ships headed to our super-stores. All in the name of feeding the god of economic growth (via the altar of hyper-consumption) in every country in the world.

The victims in all this are regular people: the workers who lose their factory jobs in Juárez and Windsor; the workers who get the factory jobs in Shenzhen and Dhaka, jobs that are by this point so degraded that some employers install nets along the perimeters of roofs to catch employees when they jump, or where safety codes are so lax that workers are killed in the hundreds when buildings collapse. The victims are also the toddlers mouthing lead-laden toys; the Walmart employee expected to work over the Thanksgiving holiday only to be trampled by a stampede of frenzied customers, while still not earning a living wage. And the Chinese villagers whose water is contaminated by one of those coal plants we use as our excuse for inaction, as well as the middle class of Beijing and Shanghai whose kids are forced to play inside because the air is so foul.[20]

A Movement Digs Its Own Grave

The greatest tragedy of all is that so much of this was eminently avoidable. We knew about the climate crisis when the rules of the new trade system were being written. After all, NAFTA was signed just one year after governments, including the United States, signed the United Nations Framework Convention on Climate Change in Rio. And it was by no means inevitable that these deals would go through. A strong coalition of North American labor and environmental groups opposed NAFTA precisely because they knew it would drive down labor and environmental standards. For a time it even looked as if they would win.

Public opinion in all three countries was deeply divided, so much so that when Bill Clinton ran for president in 1992, he pledged that he would not sign NAFTA until it substantively reflected those concerns. In Canada, Jean Chrétien campaigned for prime minister against the deal in the election of 1993. Once both were in office, however, the deal was left intact and two toothless side agreements were tacked on, one for labor and one for environmental standards. The labor movement knew better than to fall for this ploy and continued to forcefully oppose the deal, as did many Democrats in the U.S. But for a complex set of reasons having to do with a combination of reflexive political centrism and the growing influence of corporate 'partners' and donors, the leadership of many large environmental organizations decided to play ball.

'One by one, former NAFTA opponents and skeptics became enthusiastic supporters, and said so publicly,' writes journalist Mark Dowie in his critical history of the U.S. environmental movement, *Losing Ground.* These Big Green groups even created their own pro-NAFTA organization, the Environmental Coalition for NAFTA – which included the National Wildlife Federation, the Environmental Defense Fund, Conservation International, the National Audubon Society, the Natural Resources Defense Council, and the World Wildlife Fund – which, according to Dowie provided its 'unequivocal support to the agreement.' Jay Hair, then head of the National Wildlife Federation, even flew to Mexico on an official U.S. trade mission to lobby his Mexican counterparts, while attacking his critics for 'putting their protectionist polemics ahead of concern for the environment.'[21]

Not everyone in the green movement hopped on the pro-trade bandwagon: Greenpeace, Friends of the Earth, and the Sierra Club, as well as many small organizations, continued to oppose NAFTA. But that didn't matter to the Clinton administration, which had what it wanted – the ability to tell a skeptical public that 'groups representing 80 percent of national [environmental] group membership have endorsed NAFTA.' And that was important, because Clinton faced an uphill battle getting NAFTA through

Congress, with many in his own party pledging to vote against the deal. John Adams, then director of the Natural Resources Defense Council, succinctly described the extraordinarily helpful role played by groups like his: 'We broke the back of the environmental opposition to NAFTA. After we established our position Clinton only had labor to fight. We did him a big favor.'[22]

Indeed when the president signed NAFTA into law in 1993, he made a special point of thanking 'the environmental people who came out and worked through this – many of them at great criticism, particularly in the environmental movement.' Clinton also made it clear that this victory was about more than one agreement. 'Today we have the chance to do what our parents did before us. We have the opportunity to remake the world.' He explained that, 'We are on the verge of a global economic expansion . . . Already the confidence we've displayed by ratifying NAFTA has begun to bear fruit. We are now making real progress toward a worldwide trade agreement so significant that it could make the material gains of NAFTA for our country look small by comparison.' He was referring to the World Trade Organization. And just in case anyone was still worried about the environmental consequences, Clinton offered his personal assurance. 'We will seek new institutional arrangements to ensure that trade leaves the world cleaner than before.'[23]

Standing by the president's side was his vice president, Al Gore, who had been largely responsible for getting so many Big Green groups on board. Given this history, it should hardly come as a surprise that the mainstream environmental movement has been in no rush to draw attention to the disastrous climate impacts of the free trade era. To do so would only highlight their own active role in helping the U.S. government to, in Clinton's words, 'remake the world.' Much better to talk about light bulbs and fuel efficiency.

The significance of the NAFTA signing was indeed historic, tragically so. Because if the environmental movement had not been so agreeable, NAFTA might have been blocked or renegotiated to set a different kind of precedent. A new trade architecture could have been built that did not actively sabotage the fragile global climate change consensus. Instead – as had been the promise and hope of the 1992 Rio Earth Summit – this new architecture could have been grounded in the need to fight poverty and reduce emissions at the same time. So for example, trade access to developing countries could have been tied to transfers of resources and green technology so that critical new electricity and transit infrastructure was low carbon from the outset. And the deals could have been written to ensure that any measures taken to support renewable energy would not be penalized and, in fact, could be rewarded. The global economy might not have grown as quickly as it did, but it also would not be headed rapidly off the climate cliff.

The errors of this period cannot be undone, but it is not too late for a new kind of climate movement to take up the fight against so-called free trade and build this needed architecture now. That doesn't – and never did – mean an end to economic exchange across borders. It does, however, mean a far more thoughtful and deliberate approach to why we trade and whom it serves. Encouraging the frenetic and indiscriminate consumption of essentially disposable products can no longer be the system's goal. Goods must once again be made to last, and the use of energy-intensive long-haul transport will need to be rationed – reserved for those cases where goods cannot be produced locally or where local production is more carbon-intensive. (For example, growing food in greenhouses in cold parts of the United States is often more energy intensive than growing it in warmer regions and shipping it by light rail.)[24]

According to Ilana Solomon, trade analyst for the Sierra Club, this is not a fight that the climate movement can avoid. 'In order to combat climate change, there's a real need to start localizing our economies again, and thinking about how and what we're purchasing and how it's produced. And the most basic rule of trade law is you can't privilege domestic over foreign. So how do you tackle the idea of needing to incentivize local economies, tying together local green jobs policies with clean energy policies, when that is just a no-go in trade policy? . . . If we don't think about how the economy is structured, then we're actually never going to the real root of the problem.'[25]

These kinds of economic reforms would be good news – for unemployed workers, for farmers unable to compete with cheap imports, for communities that have seen their manufacturers move offshore and their local businesses replaced with big box stores. And all of these constituencies would be needed to fight for these policies, since they represent the reversal of the thirty-year trend of removing every possible limit on corporate power.

From Frenetic Expansion to Steady States

Challenging free trade orthodoxy is a heavy lift in our political culture; anything that has been in place for that long takes on an air of inevitability. But, critical as these shifts are, they are not enough to lower emissions in time. To do that, we will need to confront a logic even more entrenched than free trade – the logic of indiscriminate economic growth. This idea has understandably inspired a good deal of resistance among more liberal climate watchers, who insist that the task is merely to paint our current growth-based economic model green, so it's worth examining the numbers behind the claim.

It is Kevin Anderson of the Tyndall Centre for Climate Change Research, and one of Britain's top climate experts, who has most forcefully built the case that our growth-based economic logic is now in fundamental conflict with atmospheric limits. Addressing everyone from the U.K. Department for International Development to the Manchester City Council, Anderson has spent more than a decade patiently translating the implications of the latest climate science to politicians, economists, and campaigners. In clear and understandable language, the spiky-haired former mechanical engineer (who used to work in the petrochemical sector) lays out a rigorous road map for cutting our emissions down to a level that provides a decent shot at keeping global temperature rise below 2 degrees Celsius.

But in recent years Anderson's papers and slide shows have become more alarming. Under titles such as 'Climate Change: Going Beyond Dangerous . . .

Brutal Numbers and Tenuous Hope,' he points out that the chances of staying within anything like safe temperature levels are diminishing fast. With his colleague Alice Bows-Larkin, an atmospheric physicist and climate change mitigation expert at the Tyndall Centre, Anderson argues that we have lost so much time to political stalling and weak climate policies – all while emissions ballooned – that we are now facing cuts so drastic that they challenge the core expansionist logic at the heart of our economic system.[26]

They argue that, if the governments of developed countries want a fifty-fifty chance of hitting the agreed-upon international target of keeping warming below 2 degrees Celsius, and if reductions are to respect any kind of equity principle between rich and poor nations, then wealthy countries need to start cutting their greenhouse gas emissions by something like 8 to 10 percent a year – and they need to start right now. The idea that such deep cuts are required used to be controversial in the mainstream climate community, where the deadlines for steep reductions always seemed to be far off in the future (an 80 percent cut by 2050, for instance). But as emissions have soared and as tipping points loom, that is changing rapidly. Even Yvo de Boer, who held the U.N.'s top climate position until 2009, remarked recently that 'the only way' negotiators 'can achieve a 2-degree goal is to shut down the whole global economy.'[27]

That is a severe overstatement, yet it underlines Anderson and Bows-Larkin's point that we cannot achieve 8 to 10 percent annual cuts with the array of modest carbon-pricing or green tech solutions usually advocated by Big Green. These measures will certainly help, but they are simply not enough. That's because an 8 to 10 percent drop in emissions, year after year, is virtually unprecedented since we started powering our economies with coal. In fact, cuts above 1 percent per year 'have historically been associated only with economic recession or upheaval,' as the economist Nicholas Stern put it in his 2006 report for the British government.[28]

Even after the Soviet Union collapsed, reductions of this duration and depth did not happen (the former Soviet countries experienced average annual reductions of roughly 5 percent over a period of ten years). Nor did this level of reduction happen beyond a single-year blip after Wall Street crashed in 2008. Only in the immediate aftermath of the great market crash of 1929 did the United States see emissions drop for several consecutive years by more than 10 percent annually, but that was the worst economic crisis of modern times.[29]

If we are to avoid that kind of carnage while meeting our science-based emissions targets, carbon reduction must be managed carefully through what Anderson and Bows-Larkin describe as 'radical and immediate de-growth strategies in the US, EU and other wealthy nations.'*[30]

Now, I realize that this can all sound apocalyptic – as if reducing emissions requires economic crises that result in mass suffering. But that seems so only because we have an economic system that fetishizes GDP growth above all else, regardless of the human or ecological consequences, while failing to place value on those things that most of us cherish above all – a decent standard of living, a measure of future security, and our relationships with one another. So what Anderson and Bows-Larkin are really saying is that there is still time to avoid catastrophic warming, but not within the rules of capitalism as they are currently constructed. Which is surely the best argument there has ever been for changing those rules.[31]

Rather than pretending that we can solve the climate crisis without rocking the economic boat, Anderson

* And they don't let developing countries like China and India off the hook. According to their projections, developing countries can have just one more decade to continue to increase their emissions to aid their efforts to pull themselves out of poverty while switching over to green energy sources. By 2025, they would need to be cutting emissions 'at an unprecedented 7 per cent' a year as well.

and Bows-Larkin argue, the time has come to tell the truth, to 'liberate the science from the economics, finance and astrology, stand by the conclusions however uncomfortable . . . we need to have the audacity to think differently and conceive of alternative futures.'[32]

Interestingly, Anderson says that when he presents his radical findings in climate circles, the core facts are rarely disputed. What he hears most often are confessions from colleagues that they have simply given up hope of meeting the 2 degree temperature target, precisely because reaching it would require such a profound challenge to economic growth. 'This position is shared by many senior scientists and economists advising government,' Anderson reports.[33]

In other words, changing the earth's climate in ways that will be chaotic and disastrous is easier to accept than the prospect of changing the fundamental, growth-based, profit-seeking logic of capitalism. We probably shouldn't be surprised that some climate scientists are a little spooked by the radical implications of their own research. Most of them were quietly measuring ice cores, running global climate models, and studying ocean acidification, only to discover, as Australian climate expert and author Clive Hamilton puts it, that in breaking the news of the depth of our collective climate failure, they 'were unwittingly destabilizing the political and social order.'[34]

Nonetheless, that order has now been destabilized, which means that the rest of us are going to have to quickly figure out how to turn 'managed degrowth' into something that looks a lot less like the Great Depression and a lot more like what some innovative economic thinkers have taken to calling 'The Great Transition.'[35]

Over the past decade, many boosters of green capitalism have tried to gloss over the clashes between market logic and ecological limits by touting the wonders of green tech, or the 'decoupling' of environmental impacts from economic activity. They paint a picture of a world that can continue to function pretty much as it does now, but in which our power will come from renewable energy and all of our various gadgets and vehicles will become so much more energy-efficient that we can consume away without worrying about the impact.

If only humanity's relationship with natural resources was that simple. While it is true that renewable technologies hold tremendous promise to lower emissions, the kinds of measures that would do so on the scale we need involve building vast new electricity grids and transportation systems, often from the ground up. Even if we started construction tomorrow, it would realistically take many years, perhaps decades, before the new systems were up and running. Moreover, since we don't yet have economies powered by clean energy, all that green construction would have to burn a lot of fossil fuels in the interim – a necessary process, but one that wouldn't lower our emissions fast enough. Deep emission cuts in the wealthy nations have to start immediately. That means that if we wait for what Bows-Larkin describes as the 'whizbang technologies' to come online 'it will be too little too late.'[36]

So what to do in the meantime? Well, we do what we can. And what we can do – what doesn't require a technological and infrastructure revolution – is to consume less, right away. Policies based on encouraging people to consume less are far more difficult for our current political class to embrace than policies that are about encouraging people to consume green. Consuming green just means substituting one power source for another, or one model of consumer goods for a more efficient one. The reason we have placed all of our eggs in the green tech and green efficiency basket is precisely because these changes are safely within market logic – indeed, they encourage us to go out and buy more new, efficient, green cars and washing machines.

Consuming less, however, means changing how much energy we actually use: how often we drive, how often we fly, whether our food has to be flown to get to us, whether the goods we buy are built to last or to be replaced in two years, how large our homes are. And these are the sorts of policies that have been neglected so far. For instance, as researchers Rebecca Willis and Nick Eyre argue in a report for the U.K.'s Green Alliance, despite the fact that groceries represent roughly 12 percent of greenhouse gas emissions in Britain, 'there is virtually no government policy which is aimed at changing the way we produce, incentivising farmers for low energy farming, or how we consume, incentivising consumption of local and seasonal food.' Similarly, 'there are incentives to drive more efficient cars, but very little is done to discourage car dependent settlement patterns.'[37]

Plenty of people are attempting to change their daily lives in ways that do reduce their consumption. But if these sorts of demand-side emission reductions are to take place on anything like the scale required, they cannot be left to the lifestyle decisions of earnest urbanites who like going to farmers' markets on Saturday afternoons and wearing up-cycled clothing. We will need comprehensive policies and programs that make low-carbon choices easy and convenient for everyone. Most of all, these policies need to be fair, so that the people already struggling to cover the basics are not being asked to make additional sacrifice to offset the excess consumption of the rich. That means cheap public transit and clean light rail accessible to all; affordable, energy-efficient housing along those transit lines; cities planned for high-density living; bike lanes in which riders aren't asked to risk their lives to get to work; land management that discourages sprawl and encourages local, low-energy forms of agriculture; urban design that clusters essential services like schools and health care along transit routes and in pedestrian-friendly areas; programs that require manufacturers to be responsible for the electronic waste they produce, and to radically reduce built-in redundancies and obsolescences.*[38]

And as hundreds of millions gain access to modern energy for the first time, those who are consuming far more energy than they need would have to consume less. How much less? Climate change deniers like to claim that environmentalists want to return us to the Stone Age. The truth is that if we want to live within ecological limits, we would need to return to a

---

* A law passed by the European Parliament that would require that all cell phone manufacturers offer a common battery charger is a small step in the right direction. Similarly, requiring that electronics manufacturers use recycled metals like copper could save a great many communities from one of the most toxic mining processes in the world.

lifestyle similar to the one we had in the 1970s, before consumption levels went crazy in the 1980s. Not exactly the various forms of hardship and deprivation evoked at Heartland conferences. As Kevin Anderson explains: 'We need to give newly industrializing countries in the world the space to develop and improve the welfare and well-being of their people. This means more cuts in energy use by the developed world. It also means lifestyle changes which will have most impact on the wealthy ... We've done this in the past. In the 1960s and 1970s we enjoyed a healthy and moderate lifestyle and we need to return to this to keep emissions under control. It is a matter of the well-off 20 percent in a population taking the largest cuts. A more even society might result and we would certainly benefit from a lower carbon and more sustainable way of life.'[39]

There is no doubt that these types of policies have countless benefits besides lower emissions. They encourage civic space, physical activity, community building, as well as cleaner air and water. They also do a huge amount to reduce inequality, since it is low-income people, often people of color, who benefit most from improvements in public housing and public transit. And if strong living-wage and hire-local provisions were included in transition plans, they could also benefit most from the jobs building and running those expanded services, while becoming less dependent on jobs in dirty industries that have been disproportionately concentrated in low-income communities of color.

As Phaedra Ellis-Lamkins of the environmental justice organization Green for All puts it, 'The tools we use to combat climate change are the same tools we can use to change the game for low-income Americans and people of color . . . We need Congress to make the investments necessary to upgrade and repair our crumbling infrastructure – from building sea-walls that protect shoreline communities to fixing our storm-water systems. Doing so will create family-sustaining, local jobs. Improving our storm-water infrastructure alone would put 2 million Americans to work. We need to make sure that people of color are a part of the business community and workforce building these new systems.[40]

Another way of thinking about this is that what is needed is a fundamental reordering of the component parts of Gross Domestic Product. GDP is traditionally understood to consist of *consumption* plus *investment* plus *government spending* plus *net exports*. The free market capitalism of the past three decades has put the emphasis particularly on consumption and trade. But as we remake our economies to stay within our global carbon budget, we need to see less consumption (except among the poor), less trade (as we relocalize our economies), and less private investment in producing for excessive consumption. These reductions would be offset by increased government spending, and increased public and private investment in the infrastructure and alternatives needed to reduce our emissions to zero. Implicit in all of this is a great deal more redistribution, so that more of us can live comfortably within the planet's capacity.

Which is precisely why, when climate change deniers claim that global warming is a plot to redistribute wealth, it's not (only) because they are paranoid. It's also because they are paying attention.

Growing the Caring Economy, Shrinking the Careless One

A great deal of thought in recent years has gone into how reducing our use of material resources could be managed in ways that actually improve quality of life overall – what the French call 'selective degrowth.'*

* In French, 'decroissance' has the double meaning of challenging both growth, *croissance*, and *croire*, to believe – invoking the idea of choosing not to believe in the fiction of perpetual growth on a finite planet.

Policies like luxury taxes could be put in place to discourage wasteful consumption. The money raised could be used to support those parts of our economies that are already low-carbon and therefore do not need to contract. Obviously a huge number of jobs would be created in the sectors that are part of the green transition – in mass transit, renewable energy, weatherization, and ecosystem restoration. And those sectors that are not governed by the drive for increased yearly profit (the public sector, co-ops, local businesses, nonprofits) would expand their share of overall economic activity, as would those sectors with minimal ecological impact (such as the caregiving professions, which tend to be occupied by women and people of color and therefore underpaid). 'Expanding our economies in these directions has all sorts of advantages,' Tim Jackson, an economist at the University of Surrey and author of *Prosperity Without Growth*, has written. 'In the first place, the time spent by these professions directly improves the quality of our lives. Making them more and more efficient is not, after a certain point, actually desirable. What sense does it make to ask our teachers to teach ever bigger classes? Our doctors to treat more and more patients per hour?'[41]

There could be other benefits too, like shorter work hours, in part to create more jobs, but also because overworked people have less time to engage in low-consumption activities like gardening and cooking (because they are just too busy). Indeed, a number of researchers have analyzed the very concrete climate benefits of working less. John Stutz, a senior fellow at the Boston-based Tellus Institute, envisions that 'hours of paid work and income could converge worldwide at substantially lower levels than is seen in the developed countries today.' If countries aimed for somewhere around three to four days a week, introduced gradually over a period of decades, he argues, it could offset much of the emissions growth projected through 2030 while improving quality of life.[42]

Many degrowth and economic justice thinkers also call for the introduction of a basic annual income, a wage given to every person, regardless of income, as a recognition that the system cannot provide jobs for everyone and that it is counterproductive to force people to work in jobs that simply fuel consumption. As Alyssa Battistoni, an editor at the journal *Jacobin*, writes, 'While making people work shitty jobs to "earn" a living has always been spiteful, it's now starting to seem suicidal.'[43]

A basic income that discourages shitty work (and wasteful consumption) would also have the benefit of providing much-needed economic security in the front-line communities that are being asked to sacrifice their health so that oil companies can refine tar sands oil or gas companies can drill another fracking well. Nobody wants to have their water contaminated or have their kids suffer from asthma. But desperate people can be counted on to do desperate things – which is why we all have a vested interest in taking care of one another so that many fewer communities are faced with those impossible choices. That means rescuing the idea of a safety net that ensures that everyone has the basics covered: health care, education, food, and clean water. Indeed, fighting inequality on every front and through multiple means must be understood as a central strategy in the battle against climate change.

This kind of carefully planned economy holds out the possibility of much more humane, fulfilling lifestyles than the vast majority of us are experiencing under our current system, which is what makes the idea of a massive social movement coalescing behind such demands a real possibility. But these policies are also the most politically challenging.

Unlike encouraging energy efficiency, the measures we must take to secure a just, equitable, and inspiring transition away from fossil fuels clash directly with our reigning economic orthodoxy at every level. Such a shift breaks all the ideological rules – it requires visionary long-term planning, tough regulation of business, higher levels of taxation for the affluent, big public sector expenditure, and in many cases reversals of core privatizations in order to give communities the power to make the changes they desire. In short, it means changing everything about how we think about the economy so that our pollution doesn't change everything about our physical world.

Public and Paid For

Overcoming the Ideological Blocks to the Next Economy

When I first spotted Nastaran Mohit, she was bundled in a long puffy black coat, a white toque pulled halfway over her eyes, barking orders to volunteers gathered in an unheated warehouse. 'Take a sticky pad and write down what the needs are,' the fast-talking thirty-year-old was telling a group newly designated as Team 1. 'Okay, head on out. Who is Team 2?'[1]

It was ten days after Superstorm Sandy made landfall and we were in one of the hardest-hit neighborhoods in the Rockaways, a long, narrow strip of seaside communities in Queens, New York. The storm waters had receded but hundreds of basements were still flooded and power and cell phone service were still out. The National Guard patrolled the streets in trucks and Humvees, making sure curfew was observed, but when it came to offering help to those stranded in the cold and dark, the state and the big aid agencies were largely missing in action. (Or, more accurately, they were at the other, wealthier end of the Rockaway peninsula, where these organizations and agencies were a strong and helpful presence.)[2]

Seeing this abandonment, thousands of mostly young volunteers had organized themselves under the banner 'Occupy Sandy' (many were veterans of Occupy Wall Street) and were distributing clothes, blankets, and hot food to residents of neglected areas. They set up recovery hubs in community centers and churches, and went door-to-door in the area's notorious, towering brick housing projects, some as high as twenty-three stories. 'Muck' had become a ubiquitous verb, as in 'Do you need us to come muck out your basement?' If the answer was yes, a team of eager twenty-somethings would show up on the doorstep with mops, gloves, shovels, and bleach, ready to get the job done.

Mohit had arrived in the Rockaways to help distribute basic supplies but quickly noticed a more pressing need: in some areas, absolutely no one was providing health care. And the need was so great, it scared her. Since the 1950s, the Rockaways – once a desirable resort destination – had become a dumping ground for New York's poor and unwanted: welfare recipients, the elderly, discharged mental patients. They were crammed into high-rises, many in a part of the peninsula known locally as the 'Baghdad of Queens.'[3]

As in so many places like it, public services in the Rockaways had been cut to the bone, and then cut some more. Just six months before the storm, Peninsula Hospital Center – one of only two hospitals in the area, which served a low-income and elderly population – had shut down after the state Department of Health refused to step in. Walk-in clinics had attempted to fill the gap but they had flooded during the storm and, along with the pharmacies, had not yet reopened. 'This is just a dead-zone,' Mohit sighed.[4]

So she and friends in Occupy Sandy called all the doctors and nurses they knew and asked them to bring in whatever supplies they could. Next, they convinced the owner of an old furrier, damaged in the storm, to let them convert his storefront on the neighborhood's main drag into a makeshift MASH unit. There, amidst the animal pelts hanging from the ceiling, volunteer doctors and nurses began to see patients, treat wounds, write prescriptions, and provide trauma counseling.

There was no shortage of patients; in its first two weeks, Mohit estimated that the clinic helped hundreds of people. But on the day I visited, worries were mounting about the people still stuck in the highrises. As volunteers went door-to-door distributing supplies in the darkened projects, flashlights strapped to their foreheads, they were finding alarming numbers of sick people. Cancer and HIV/AIDS meds had run out, oxygen tanks were empty, diabetics were out of insulin, and addicts were in withdrawal. Some people were too sick to brave the dark stairwells and multiple flights of stairs to get help; some didn't leave because they had nowhere to go and no way to get off the peninsula (subways and buses were not operating); others feared that if they left their apartments, their homes would be burglarized. And without cell service or power for their TVs, many had no idea what was going on outside.

Most shockingly, residents reported that until Occupy Sandy showed up, no one had knocked on their doors since the storm. Not from the Health Department, nor the city Housing Authority (responsible for running the projects), nor the big relief agencies like the Red Cross. 'I was like "Holy crap,"' Mohit told me. 'There was just no medical attention at all.'[*5] Referring to the legendary abandonment of

* This was the situation not only in the Rockaways but seemingly wherever public housing was in the path of the storm. In Red Hook, Brooklyn, many residents were left without power for three weeks, during which time the Housing Authority never went systematically door-to-door. As sixty-year-old Wally Bazemore put it at an angry residents meeting: 'We were literally in the dark and we were completely in the dark.'

New Orleans's poor residents when the city flooded in 2005, she said: 'This is Katrina 2.0.'[6]

The most frustrating part was that even when a pressing health need was identified, and even when the volunteer doctors wrote the required prescriptions, 'we bring it to the pharmacy and the pharmacy is sending it back to us because they need insurance information. And then we get as much information as we can and we bring it back and they say, "Now we need their Social Security number." '[7]

According to a 2009 Harvard Medical School study, as many as 45,000 people die annually in the United States because they lack health insurance. As one of the study's co-authors pointed out, this works out to about one death every twelve minutes. President Obama's stunted 2010 health care law may have somewhat improved those numbers, but watching the insurance companies continue to put money before human health in the midst of the worst storm in New York's history cast this preexisting injustice in a new, more urgent light. 'We need universal health care,' Mohit declared. 'There is no other way around it. There is absolutely no other way around it.' Anyone who disagreed should come to the disaster zone, she said, because this 'is a perfect situation for people to really examine how nonsensical, inhumane, and barbaric this system is.'[8]

The word 'apocalypse' derives from the Greek *apokalypsis*, which means 'something uncovered' or revealed. Besides the need for a dramatically better health care system, there was much else uncovered and revealed when the floodwaters retreated in New York that October. The disaster revealed how dangerous it is to be dependent on centralized forms of energy that can be knocked out in one blow. It revealed the life-and-death cost of social isolation, since it was the people who did not know their neighbors, or who were frightened of them, who were most at risk. Meanwhile, it was the tightest-knit communities, where neighbors took responsibility for one another's safety, that were best able to literally weather the storm.

The disaster also revealed the huge risks that come with deep inequality, since the people who were already the most vulnerable – undocumented workers, the formerly incarcerated, people in public housing – suffered most and longest. In low-income neighborhoods, homes filled not only with water but with heavy chemicals and detergents – the legacy of systemic environmental racism that allowed toxic industries to build in areas inhabited mostly by people of color. Public housing projects that had been left to decay – while the city bided its time before selling them off to developers – turned into death traps, their ancient plumbing and electrical systems giving way completely. As Aria Doe, executive director of the Action Center for Education and Community Development in the Rockaways, put it, the peninsula's poorest residents 'were six feet under' before the storm even hit. 'Right now, they're seven or eight feet under.'[9]

All around the world, the hard realities of a warming world are crashing up against the brutal logic of austerity, revealing just how untenable it is to starve the public sphere at the very moment we need it most. The floods that hit the U.K. in the winter of 2013–2014, for instance, would have been trying for any government: thousands of homes and workplaces were inundated, hundreds of thousands of houses and other buildings lost power, farmland was submerged, several rail lines were down for weeks, all combining to create what one top official called an 'almost unparalleled natural disaster.' This as the country was still reeling from a previous devastating storm that had struck just two months before.[10]

But the floods were particularly awkward for the coalition government led by Conservative prime minister David Cameron because, in the three years prior, it had gutted the Environment Agency (EA), which was responsible for dealing with flooding. Since 2009, at least 1,150 jobs had been lost at the agency, with as many as 1,700 more on the chopping block, adding up to approximately a quarter of its total workforce. In 2012 *The Guardian* had revealed that 'nearly 300 flood defence schemes across England [had] been left unbuilt due to government budget cuts.' The head of the Environment Agency had stated plainly during the most recent round of cuts that 'Flood risk maintenance will be impacted.'[11]

Cameron is no climate change denier, which is what made it all the more incredible that he had hobbled the agency responsible for protecting the public from rising waters and more ferocious storms, two well-understood impacts of climate change. And his praise of the good works of the staff that had survived his axe provided cold comfort. 'It is a disgrace that the Government is happy to put cost cutting before public safety and protecting family homes,' announced the trade union representing EA workers in a scathing statement. 'They can't have it both ways, praising the sterling work of members in the Agency in one breath, and in the next breath announcing further damaging cuts.'[12]

During good times, it's easy to deride 'big government' and talk about the inevitability of cutbacks. But during disasters, most everyone loses their free market religion and wants to know that their government has their backs. And if there is one thing we can be sure of, it's that extreme weather events like Superstorm Sandy, Typhoon Haiyan in the Philippines, and the British floods – disasters that, combined, pummeled coastlines beyond recognition, ravaged millions of homes, and killed many thousands – are going to keep coming.

Over the course of the 1970s, there were 660 reported disasters around the world, including droughts, floods, extreme temperature events, wildfires, and storms. In the 2000s, there were 3,322 – a fivefold boost. That is a staggering increase in just over thirty years, and clearly global warming cannot be said to have 'caused' all of it. But the climate signal is also clear. 'There's no question that climate change has increased the frequency of certain types of extreme weather events,' climate scientist Michael Mann told me in an interview, 'including drought, intense hurricanes, and super typhoons, the frequency and intensity and duration of heat waves, and potentially other types of extreme weather though the details are still being debated within the scientific community.'[13]

Yet these are the same three decades in which almost every government in the world has been steadily chipping away at the health and resilience of the public sphere. And it is this neglect that, over and over again, turns natural disasters into unnatural catastrophes. Storms burst through neglected levees. Heavy rain causes decrepit sewer systems to back up and overflow. Wildfires rage out of control for lack of workers and equipment to fight them (in Greece, fire departments can't afford spare tires for their trucks driving into forest blazes). Emergency responders are missing in action for days after a major hurricane. Bridges and tunnels, left in a state of disrepair, collapse under the added pressure.

The costs of coping with increasing weather extremes are astronomical. In the United States, each major disaster seems to cost taxpayers upward of a billion dollars. The cost of Superstorm Sandy is estimated at $65 billion. And that was just one year after Hurricane Irene caused around $10 billion in damage, just one episode in a year that saw fourteen billion-dollar disasters in the U.S. alone. At that time, 2011 held the title as the costliest year ever for disasters, with total damages reaching at least $380 billion. And with policymakers still locked in the vise grip of austerity logic, these rising emergency expenditures are being offset with cuts to everyday public spending, which will make societies even more vulnerable during the next disaster – a classic vicious cycle.[14]

It was never a good idea to neglect the foundations of our societies in this way. In the context of climate change, however, that decision looks suicidal. There are many important debates to be had about the best way to respond to climate change – storm walls or ecosystem restoration? Decentralized renewables, industrial scale wind power combined with natural gas, or nuclear power? Small-scale organic farms or industrial food systems? There is, however, *no* scenario in which we can avoid wartime levels of spending in the public sector – not if we are serious about preventing catastrophic levels of warming, and minimizing the destructive potential of the coming storms.

It's no mystery where that public money needs to be spent. Much of it should go to the kinds of ambitious emission-reducing projects already discussed – the smart grids, the light rail, the citywide composting systems, the building retrofits, the visionary transit systems, the urban redesigns to keep us from spending half our lives in traffic jams. The private sector is ill suited to taking on most of these large infrastructure investments: if the services are to be accessible, which they must be in order to be effective, the profit margins that attract private players simply aren't there.

Transit is a good example. In March 2014, when air pollution in French cities reached dangerously high levels, officials in Paris made a snap decision to discourage car use by making public transit free for three days. Obviously private operators would strenuously resist such measures. And yet by all rights, our transit systems should be responding with the same kind of urgency to dangerously high levels of atmospheric carbon. Rather than allowing subway and bus fares to rise while service erodes, we need to be lowering prices and expanding services – regardless of the costs.

Public dollars also need to go to the equally important, though less glamorous projects and services that will help us prepare for the coming heavy weather. That includes things like hiring more firefighters and improving storm barriers. And it means coming up with new, nonprofit disaster insurance programs so that people who have lost everything to a hurricane or a forest fire are not left at the mercy of a private insurance industry that is already adapting to climate change by avoiding payouts and slapping victims with massive rate increases. According to Amy Bach, cofounder of the San Francisco-based advocacy group United Policyholders, disaster insurance is becoming 'very much like health insurance. We're going to have to increasingly take the profit motive out of the system so that it operates efficiently and effectively, but without generating obscene executive salaries and bonuses and shareholder returns. Because it's not going to be a sustainable model. A publicly traded insurance company in the face of climate change is not a sustainable business model for the end user, the consumer.'[15] It's that or a disaster capitalism free-for-all; those are the choices.

These types of improvements are of course in far greater demand in developing countries like the Philippines, Kenya, and Bangladesh that are already facing some of the most severe climate impacts. Hundreds of billions of dollars are urgently needed to build seawalls; storage and distribution networks for food, water, and medicine; early warning systems and shelters for hurricanes, cyclones, and tsunamis – as well as public health systems able to cope with increases in climate-related diseases like malaria.[16] Though mechanisms to protect against government corruption are needed, these countries should not have to spend their health care and education budgets on costly disaster insurance plans purchased from transnational corporations, as is happening right now. Their people should be receiving direct compensation from the countries (and companies) most responsible for warming the planet.

The Polluter Pays

About now a sensible reader would be asking: how on earth are we going to pay for all this? It's the essential question. A 2011 survey by the U.N. Department of

Economic and Social Affairs looked at how much it would cost for humanity to 'overcome poverty, increase food production to eradicate hunger without degrading land and water resources, and avert the climate change catastrophe.' The price tag was $1.9 trillion a year for the next forty years – and 'at least one half of the required investments would have to be realized in developing countries.'[17]

As we all know, public spending is going in the opposite direction almost everywhere except for a handful of fast-growing so-called emerging economies. In North America and Europe, the economic crisis that began in 2008 is still being used as a pretext to slash aid abroad and cut climate programs at home. All over Southern Europe, environmental policies and regulations have been clawed back, most tragically in Spain, which, facing fierce austerity pressure, drastically cut subsidies for renewables projects, sending solar projects and wind farms spiraling toward default and closure. The U.K. under David Cameron also cut supports for renewable energy.

So if we accept that governments are broke, and they're not likely to introduce 'quantitative easing' (aka printing money) for the climate system as they have for the banks, where is the money supposed to come from? Since we have only a few short years to dramatically lower our emissions, the only rational way forward is to fully embrace the principle already well established in Western law: the polluter pays.

The fossil fuel companies have known for decades that their core product was warming the planet, and yet they have not only failed to adapt to that reality, they have actively blocked progress at every turn. Meanwhile, oil and gas companies remain some of the most profitable corporations in history, with the top five oil companies pulling in $900 billion in profits from 2001 to 2010. ExxonMobil still holds the record for the highest corporate profits ever reported in the United States, earning $41 billion in 2011 and $45 billion in 2012. These companies are rich, quite simply, because they have dumped the cost of cleaning up their mess onto regular people around the world. It is this situation that, most fundamentally, needs to change.[18]

And it will not change without strong action. For well over a decade, several of the oil majors have claimed to be voluntarily using their profits to invest in a shift to renewable energy. In 2000, BP rebranded itself 'Beyond Petroleum' and even changed its logo to a sunburst, called 'the Helios mark after the sun god of ancient Greece.' ('We are not an oil company,' then-chief executive Sir John Browne said at the time, explaining that, 'We are aware the world wants less carbon-intensive fuels. What we want to do is create options.') Chevron, for its part, ran a high-profile advertising campaign declaring, 'It's time oil companies get behind renewables ... We agree.' But according to a study by the Center for American Progress, just 4 percent of the Big Five's $100 billion in combined profits in 2008 went to 'renewable and alternative energy ventures.' Instead, they continue to pour their profits into shareholder pockets, outrageous executive pay (Exxon CEO Rex Tillerson makes more than $100,000 a day), and new technologies designed to extract even dirtier and more dangerous fossil fuels.[19]

And even as the demand for renewables increases, the percentage the fossil fuel companies spend on them keeps shrinking – by 2011, most of the majors were spending less than 1 percent of their overall expenditures on alternative energy, with Chevron and Shell spending a deeply unimpressive 2.5 percent. In 2014, Chevron pulled back even further. According to *Bloomberg Businessweek*, the staff of a renewables division that had almost doubled its target profits was told 'that funding for the effort would dry up' and was urged 'to find jobs elsewhere.' Chevron also moved to sell off businesses that had developed green projects for governments and school districts. As oil industry watcher Antonia Juhasz has observed, 'You wouldn't know it from their advertising, but the world's major oil companies have either entirely divested from alternative energy or significantly reduced their investments in favor of doubling down on ever-more risky and destructive sources of oil and natural gas.'[20]

Given this track record, it's safe to assume that if fossil fuel companies are going to help pay for the shift to renewable energy, and for the broader costs of a climate destabilized by their pollution, it will be because they are forced to do so by law. Just as tobacco companies have been obliged to pay the costs of help-ing people to quit smoking, and BP has had to pay for much of the cleanup of its oil spill in the Gulf of Mex-ico, it is high time for the industry to at least split the bill for the climate crisis. And there is mounting evi-dence that the financial world understands that this is coming. In its 2013 annual report on 'Global Risks,' the World Economic Forum (host of the annual superelite gathering in Davos) stated plainly, 'Although the Alaskan village of Kivalina – which faces being "wiped out" by the changing climate – was unsuccessful in its attempts to file a US$ 400 million lawsuit against oil and coal companies, future plaintiffs may be more successful. Five decades ago, the U.S. tobacco indus-try would not have suspected that in 1997 it would agree to pay $368 billion in health-related damages.' But it did.[21]

The question is: how do we stop fossil fuel profits

from continuing to hemorrhage into executive pay-checks and shareholder pockets – and how do we do it soon, before the companies are significantly less profit-able or out of business because we have moved to a new energy system? As the Global Risks report suggests, communities severely impacted by climate change have made several attempts to use the courts to sue for damages, but so far they have been unsuccessful. A steep carbon tax would be a straightforward way to get a piece of the profits, as long as it contained a generous redistributive mechanism – a tax cut or income credit – that compensated poor and middle-class consumers for increased fuel and heating prices. As Canadian econo-mist Marc Lee points out, designed properly, 'It is possible to have a progressive carbon tax system that reduces inequality as it raises the price of emitting greenhouse gases.'[22] An even more direct route to get-ting a piece of those pollution profits would be for governments to negotiate much higher royalty rates on oil, gas, and coal extraction, with the revenues going to 'heritage trust funds' that would be dedicated to build-ing the post-fossil fuel future, as well as to helping communities and workers adapt to these new realities.

Fossil fuel corporations can be counted on to resist any new rules that cut into their profits, so harsh penal-ties, including revoking corporate charters, would need to be on the table. Companies would threaten to pull out of certain operations, to be sure, but once a multinational like Shell has spent billions to build the mines and drilling platforms needed to extract fossil fuels, it is unlikely to abandon that infrastructure because royalties go up. (Though it will bitterly complain and may well seek damages at an investment tribunal.)

But the extractive industries shouldn't be the only targets of the 'polluter pays' principle. The U.S. military is by some accounts the largest single consumer of petroleum in the world. In 2011, the Department of Defense released, at minimum, 56.6 million metric tons of CO_2 equivalent into the atmosphere, more than the U.S.-based operations of ExxonMobil and Shell combined.[23] So surely the arms companies should pay their share. The car companies have plenty to answer for too, as do the shipping industry and the airlines.

Moreover, there is a simple, direct correlation between wealth and emissions – more money generally means more flying, driving, boating, and powering of multiple homes. One case study of German consumers indicates that the travel habits of the most affluent class have an impact on climate 250 percent greater than that of their lowest-earning neighbors.[24]

That means any attempt to tax the extraordinary concentration of wealth at the very top of the economic pyramid, as documented so persuasively by Thomas

Piketty among many others, would – if partially chan-
neled into climate financing – effectively make the
polluters pay. As journalist and climate and energy policy expert Gar Lipow puts it, 'We should tax the rich
more because it is the fair thing to do, and because it
will provide a better life for most of us, and a more
prosperous economy. However, providing money to
save civilization and reduce the risk of human extinc-
tion is another good reason to bill the rich for their fair
share of taxes.' But it must be said that a 'polluter pays'
principle would have to reach beyond the super rich.
According to Stephen Pacala, director of the Princeton
Environmental Institute and codirector of Princeton's
Carbon Mitigation Initiative, the roughly 500 million
richest of us on the planet are responsible for about half
of all global emissions. That would include the rich in
every country in the world, notably in countries like
China and India, as well as significant parts of the middle classes in North America and Europe.*[25]

Taken together, there is no shortage of options for

* This is why the persistent positing of population control
as a solution to climate change is a distraction and moral
dead end. As this research makes clear, the most significant cause of rising emissions is not the reproductive
behavior of the poor but the consumer behaviors of the
rich.

equitably coming up with the cash to prepare for the coming storms while radically lowering our emissions to prevent catastrophic warming.

Consider the following list, by no means complete:

- A 'low-rate' financial transaction tax – which would hit trades of stocks, derivatives, and other financial instruments – could bring in nearly $650 billion at the global level each year, according to a 2011 resolution of the European Parliament (and it would have the added bonus of slowing down financial speculation).[26]

- Closing tax havens would yield another windfall. The U.K.-based Tax Justice Network estimates that in 2010, the private financial wealth of individuals stowed unreported in tax havens around the globe was somewhere between $21 trillion and $32 trillion. If that money were brought into the light and its earnings taxed at a 30 percent rate, it would yield at least $190 billion in income tax revenue each year.[27]

- A 1 percent 'billionaire's tax,' floated by the U.N., could raise $46 billion annually.[28]

- Slashing the military budgets of each of the top ten military spenders by 25 percent could free up another $325 billion, using 2012 numbers reported by the Stockholm International Peace Research Institute. (Granted, probably the toughest sell of all, particularly in the U.S.)[29]

- A $50 tax per metric ton of CO_2 emitted in developed countries would raise an estimated $450 billion annually, while a more modest $25 carbon tax would still yield $250 billion per year, according to a 2011 report by the World Bank, the International Monetary Fund, and the Organisation for Economic Co-operation and Development (OECD), among others.[30]
- Phasing out fossil fuel subsidies globally would conservatively save governments a total $775 billion in a single year, according to a 2012 estimate by Oil Change International and the Natural Resources Defense Council.[31]

If these various measures were taken together, they would raise more than $2 trillion annually.[32] Certainly enough for a very healthy start to finance a Great Transition (and avoid a Great Depression). And that doesn't count any royalty increases on fossil fuel extraction. Of course, for any of these tax crackdowns to work, key governments would have to coordinate their responses so that corporations had nowhere to hide – a difficult task, though far from impossible, and one frequently bandied about at G20 summits.

In addition to the simple fact that the money is badly needed, there are practical political reasons why 'polluter pays' should guide climate financing. As we have seen, responding to the climate crisis can offer real benefits to a majority of people, but real solutions will also, by definition, require short- and medium-term sacrifices and inconveniences. And what we know from past sacrifices made in the name of a crisis – most notably via rationing, conservation, and price controls during both world wars – is that success depends entirely on a perception of fairness.

In Britain and North America during World War II, for instance, every strata of society was required to make do with less, even the very rich. And in fact, though overall consumption in the U.K. dropped by 16 percent, caloric intake for the poor increased during the war, because the rations provided low-income people with more than they could otherwise afford.[33]

There was plenty of cheating and black market profiteering, of course, but these programs enjoyed broad-based support because they were, at least in theory, fair. The theme of equality pervaded government campaigns about these wartime programs: 'Fair Shares for All' was a key slogan in the U.K, while the U.S. went with 'Share and Share Alike' and 'Produce,

Conserve, Share and Play Square.'[34] An Office of Price Administration pamphlet from 1942 argued that rationing was part of the American tradition. 'What Is Rationing?' it asked.

> First, let's be sure what rationing is not. It is not star-vation, long bread lines, shoddy goods. Rather, it is a community plan for dividing fairly the supplies we have among all who need them. Second, it is not 'un-American.' The earliest settlers of this country, facing scarcities of food and clothing, pooled their precious supplies and apportioned them out to everyone on an equal basis. It was an American idea then, and it is an American idea now, to share and share alike – to sac-rifice, when necessary, but sacrifice together, when the country's welfare demands it.[35]

Governments also made sure that there were very public crackdowns on wealthy and well-connected individuals who broke the rules, sending the message that no one was exempt. In the U.K., movie stars, as well as corporations like Woolworth and Sainsbury, faced prosecution for rations violations. In the United States, cases were brought against some of the largest corporations in the country. It was no secret that many large U.S. manufacturers disliked the entire rationing system; they lobbied against it, because they believed it eroded their brand value. Yet they were forced to accept it all the same.[36]

This perception of fairness – that one set of rules applies to players big and small – has been entirely missing from our collective responses to climate change thus far. For decades, regular people have been asked to turn off their lights, put on sweaters, and pay premium prices for nontoxic cleaning products and renewable energy – and then watched as the biggest polluters have been allowed to expand their emissions without penalty. This has been the pattern ever since President Jimmy Carter addressed the American public in July 1979 about the fact that 'too many of us now tend to worship self-indulgence and consumption. Human identity is no longer defined by what one does, but by what one owns.' He urged Americans 'for your good and for your nation's security to take no unnecessary trips, to use carpools or public transportation whenever you can, to park your car one extra day per week, to obey the speed limit, and to set your thermostats to save fuel. Every act of energy conservation like this is more than just common sense – I tell you it is an act of patriotism.'[37]

The address was initially well received but came to be derided as the 'malaise' speech and is frequently cited as one of the reasons Carter lost his reelection bid to Ronald Reagan. And though he was not talking about climate change but rather a broad 'crisis of confidence' against a backdrop of energy scarcity, the speech is still invoked as proof that any politician who asks voters to sacrifice to solve an environmental crisis is on a suicide mission. Indeed this assessment has shaped the win-win messaging of environmentalists ever since.

So it's interesting to note that the late intellectual Christopher Lasch, who was one of Carter's key advisors on the infamous speech, was also one of its most pointed critics. The author of *The Culture of Narcissism* had strongly urged the president to temper his message of personal austerity with assurances of fundamental fairness and social justice. As Lasch revealed to an interviewer years later, he had told Carter to 'put a more populist construction in his indictment of American consumerism . . . What was needed was a program that called for sacrifices all right, but made it clear that the sacrifices would be distributed in an equitable fashion.' And that, Lasch said, 'would mean that those most able to make sacrifices would be the ones on whom the sacrifices fell. That's what I mean by populism.'[38]

We cannot know if the reaction might have differed had Carter listened to that advice and presented a plan for conservation that began with those pushing and profiting most from wasteful consumption. We do know that responses to climate change that continue to put the entire burden on individual consumers are doomed to fail. For instance, the annual 'British Social Attitudes' survey, conducted by the independent NatCen Social Research, asked a set of questions about climate policies in the year 2000, and then again in 2010. It found that, 'Whereas, 43 per cent a decade ago said they would be willing to pay higher prices to protect the environment, this is nowadays only true of 26 per cent. There has been a similar fall in the proportion prepared to pay higher taxes (31 to 22 per cent), but a smaller decline in relation to cuts in the standard of living (26 per cent to 20 per cent).'[39]

These results, and others like them, have been cited as proof that during times of economic hardship, people's environmental concerns go out the window. But that is not what these polls prove. Yes, there has been a drop in the willingness of individuals to bear the financial burden of responding to climate change, but not simply because economic times are hard. Western governments have responded to these hard times – which have been created by rampant greed and corruption among their wealthiest citizens – by asking those least responsible for the current conditions to bear the burden. After paying for the crisis of the bankers with cuts to education, health care, and social safety nets, is it any wonder that a beleaguered public is in no mood to bail out the fossil fuel companies from the crisis that they not only created but continue to actively worsen?

Most of these surveys, notably, don't ask respondents how they feel about raising taxes on the rich and removing fossil fuel subsidies, yet these are some of the most reliably popular policies around. And it's worth noting that a U.S. poll conducted in 2010 – with the country still reeling from economic crisis – asked voters whether they would support a plan that 'would make oil and coal companies pay for the pollution they cause. It would encourage the creation of new jobs and new technologies in cleaner energy like wind, solar, and nuclear power. The proposal also aims to protect working families, so it refunds almost all of the money it collects directly to the American people, like a tax refund, and most families end up better off.' The poll found that three quarters of voters, including the vast majority of Republicans, supported the ideas as outlined, and only 11 percent strongly opposed it. The plan was similar to a proposal, known as 'cap and dividend,' being floated by a pair of senators at the time, but it was never seriously considered by the U.S. Senate.⁴⁰

And when, in June 2014, Obama finally introduced plans to use the Environmental Protection Agency to limit greenhouse gas emissions from existing power plants, the coal lobby howled with indignation but public opinion was solidly supportive. According to one poll, 64 percent of Americans, including a great many Republicans, backed such a policy even though it would likely mean paying more for energy every month.[41]

The lesson from all this is not that people won't sacrifice in the face of the climate crisis. It's that they have had it with our culture of *lopsided* sacrifice, in which individuals are asked to pay higher prices for supposedly green choices while large corporations dodge regulation and not only refuse to change their behavior, but charge ahead with ever more polluting activities. Witnessing this, it is perfectly sensible for people to shed much of the keener enthusiasm that marked the early days of the climate movement, and to make it clear that no more sacrifice will be made until the policy solutions on the table are perceived as just. This does not mean the middle class is off the hook. To fund the kind of social programs that will make a just transition possible, taxes will have to rise for everyone but the poor. But if the funds raised go toward social programs and services that reduce inequality and make lives far less insecure and precarious, then public attitudes toward taxation would very likely shift as well.

To state the obvious: it would be incredibly difficult to persuade governments in almost every country in the world to implement the kinds of redistributive climate mechanisms I have outlined. But we should be clear about the nature of the challenge: it is not that 'we' are broke or that we lack options. It is that our political class is utterly unwilling to go where the money is (unless it's for a campaign contribution), and the corporate class is dead set against paying its fair share.

Seen in this light, it's hardly surprising that our leaders have so far failed to act to avert climate chaos. Indeed even if aggressive 'polluter pays' measures were introduced, it isn't at all clear that the current political class would know what to do with the money. After all, changing the building blocks of our societies – the energy that powers our economies, how we move around, the designs of our major cities – is not about writing a few checks. It requires bold long-term planning at every level of government, and a willingness to stand up to polluters whose actions put us all in danger. And that won't happen until the corporate liberation project that has shaped our political culture for three and a half decades is buried for good.

There is a direct relationship between breaking fossilized free market rules and making swift progress on climate change. Which is why, if we are to collectively meet the enormous challenges of this crisis, a

robust social movement will need to demand (and create) political leadership that is not only committed to
making polluters pay for a climate-ready public sphere,
but willing to revive two lost arts: long-term public
planning, and saying no to powerful corporations.

The Leap Years

Just Enough Time for Impossible

The crucial question we are left with, then, is this: has an economic shift of this kind *ever* happened before in history? We know it can happen during wartime, when presidents and prime ministers are the ones commanding the transformation from above. But has it ever been demanded from below, by regular people, when leaders have wholly abdicated their responsibilities? Having combed through the history of social movements in search of precedents, I must report that the answer to that question is predictably complex, filled with 'sort ofs' and 'almosts' – but also at least one 'yes.'

In the West, the most common precedents invoked to show that social movements really can be a disruptive historical force are the celebrated human rights movements of the past century – most prominently, civil, women's, and gay and lesbian rights. And these movements unquestionably transformed the face and texture of the dominant culture. But given that the challenge for the climate movement hinges on pulling off a profound and radical *economic* transformation, it must be noted that for these movements, the legal and cultural battles were always more successful than the economic ones.

The U.S. civil rights movement, for instance, fought not only against legalized segregation and discrimination but also *for* massive investments in schools and jobs programs that would close the economic gap between blacks and whites once and for all. In his 1967 book, *Where Do We Go from Here: Chaos or Community?*, Martin Luther King Jr pointed out that, 'The practical cost of change for the nation up to this point has been cheap. The limited reforms have been obtained at bargain rates. There are no expenses, and no taxes are required, for Negroes to share lunch counters, libraries, parks, hotels and other facilities with whites ... The real cost lies ahead ... The discount education given Negroes will in the future have to be purchased at full price if quality education is to be realized. Jobs are harder and costlier to create than voting rolls. The eradication of slums housing millions is complex far beyond integrating buses and lunch counters.'[1]

And though often forgotten, the more radical wing of the second-wave feminist movement also argued for fundamental challenges to the free market economic order. It wanted women not only to get equal pay for equal work in traditional jobs but to have their work in the home caring for children and the elderly recognized and compensated as a massive unacknowledged market subsidy – essentially a demand for wealth redistribution on a scale greater than the New Deal.

But as we know, while these movements won huge battles against institutional discrimination, the victories that remained elusive were those that, in King's words, could not be purchased 'at bargain rates.' There would be no massive investments in jobs, schools, and decent homes for African Americans, just as the 1970s women's movement would not win its demand for 'wages for housework' (indeed paid maternity leave remains a battle in large parts of the world). Sharing legal status is one thing; sharing resources quite another.

If there is an exception to this rule it is the huge gains won by the labor movement in the aftermath of the Great Depression – the massive wave of unionization that forced owners to share a great deal more wealth with their workers, which in turn helped create a context to demand ambitious social programs like Social Security and unemployment insurance (programs from which the majority of African American and many women workers were notably excluded). And in response to the market crash of 1929, tough new rules regulating the financial sector were introduced at real cost to unfettered profit making. In the same period, social movement pressure created the conditions for the New Deal and programs like it across the industrialized world. These made massive investments in public infrastructure – utilities, transportation systems, housing, and more – on a scale comparable to what the climate crisis calls for today.

If the search for historical precedents is extended more globally (an impossibly large task in this context, but worth a try), then the lessons are similarly mixed. Since the 1950s, several democratically elected socialist governments have nationalized large parts of their extractive sectors and begun to redistribute to the poor and middle class the wealth that had previously hemorrhaged into foreign bank accounts, most notably Mohammad Mosaddegh in Iran and Salvador Allende in Chile. But those experiments were interrupted by foreign-sponsored coups d'état before reaching their potential. Indeed postcolonial independence movements – which so often had the redistribution of unjustly concentrated resources, whether of land or minerals, as their core missions – were consistently undermined through political assassinations, foreign interference, and, more recently, the chains of debt-driven structural adjustment programs (not to mention the corruption of local elites).

Even the stunningly successful battle against

apartheid in South Africa suffered its most significant losses on the economic equality front. The country's freedom fighters were not, it is worth remembering, only demanding the right to vote and move freely. They were also, as the African National Congress's official policy platform, the Freedom Charter, made clear, struggling for key sectors of the economy – including the mines and the banks – to be nationalized, with their proceeds used to pay for the social programs that would lift millions in the townships out of poverty. Black South Africans won their core legal and electoral battles, but the wealth accumulated under apartheid remained intact, with poverty deepening significantly in the post-apartheid era.[2]

There have been social movements, however, that have succeeded in challenging entrenched wealth in ways that are comparable to what today's movements must provoke if we are to avert climate catastrophe. These are the movements for the abolition of slavery and for Third World independence from colonial powers. Both of these transformative movements forced ruling elites to relinquish practices that were still extraordinarily profitable, much as fossil fuel extraction is today.

The movement for the abolition of slavery in particular shows us that a transition as large as the one confronting us today has happened before – and indeed it is remembered as one of the greatest moments in human history. The economic impacts of slavery abolition in the mid-nineteenth century have some striking parallels with the impacts of radical emission reduction, as several historians and commentators have observed. Journalist and broadcaster Chris Hayes, in an award-winning 2014 essay titled 'The New Abolitionism,' pointed out 'the climate justice movement is demanding that an existing set of political and economic interests be forced to say good-bye to trillions of dollars of wealth' and concluded that 'it is impossible to point to any precedent other than abolition.'[3]

There is no question that for a large sector of the ruling class at the time, losing the legal right to exploit men and women in bondage represented a major economic blow, one as huge as the one players ranging from Exxon to Richard Branson would have to take today. As the historian Greg Grandin has put it, 'In the realm of economics, the importance of slaves went well beyond the wealth generated from their uncompensated labor. Slavery was the flywheel on which America's market revolution turned – not just in the United States, but in all of the Americas.' In the eighteenth century, Caribbean sugar plantations, which were wholly dependent on slave labor, were by far the most profitable outposts of the British Empire, generating revenues that far outstripped the other colonies. In *Bury the Chains*, Adam Hochschild quotes enthusiastic slave traders describing the buying and selling of humans as 'the hinge on which all the trade of this globe moves' and 'the foundation of our commerce . . . and first cause of our national industry and riches.'[4]

While not equivalent, the dependency of the U.S. economy on slave labor – particularly in the Southern states – is certainly comparable to the modern global economy's reliance on fossil fuels.* According to historian Eric Foner, at the start of the Civil War, 'slaves as property were worth more than all the banks, factories and railroads in the country put together.' Strengthening the parallel with fossil fuels, Hayes points out that 'in 1860, slaves represented about 16 percent of the total household assets – that is, all the

---

* The reliance was certainly not limited to the Southern states: cutting-edge historical research has been exploding long-held perceptions that the North and South of the United States had distinct and irreconcilable economies in this period. In fact, Northern industrialists and Wall Street were far more dependent on and connected to slavery than has often been assumed, and even some crucial innovations in scientific management and accounting can be traced to the American plantation economy.

wealth – in the entire [United States], which in today's terms is a stunning $10 trillion.' That figure is very roughly similar to the value of the carbon reserves that must be left in the ground worldwide if we are to have a good chance of keeping warming below 2 degrees Celsius.[5]

But the analogy, as all acknowledge, is far from perfect. Burning fossil fuels is of course not the moral equivalent of owning slaves or occupying countries. (Though heading an oil company that actively sabotages climate science, lobbies aggressively against emission controls while laying claim to enough interred carbon to drown populous nations like Bangladesh and boil sub-Saharan Africa is indeed a heinous moral crime.) Nor were the movements that ended slavery and defeated colonial rule in any way bloodless: nonviolent tactics like boycotts and protests played major roles, but slavery in the Caribbean was only outlawed after numerous slave rebellions were brutally suppressed, and, of course, abolition in the United States came only after the carnage of the Civil War.

Another problem with the analogy is that, though the liberation of millions of slaves in this period – some 800,000 in the British colonies and four million in the U.S. – represents the greatest human rights victory of its time (or, arguably, any time), the economic side of the struggle was far less successful. Local and international elites often managed to extract steep payoffs to compensate themselves for their 'losses' of human property, while offering little or nothing to former slaves. Washington broke its promise, made near the end of the Civil War, to grant freed slaves ownership of large swaths of land in the U.S. South (a pledge known colloquially as '40 acres and a mule'). Instead the lands were returned to former slave owners, who proceeded to staff them through the indentured servitude of sharecropping. Britain, as discussed, awarded massive paydays to its slave owners at the time of abolition. And France, most shockingly, sent a flotilla of warships to demand that the newly liberated nation of Haiti pay a huge sum to the French crown for the loss of its bonded workforce – or face attack.[6] Reparations, but in reverse.

The costs of these, and so many other gruesomely unjust extortions, are still being paid in lives, from Haiti to Mozambique. The reverse-reparations saddled newly liberated nations and people with odious debts that deprived them of true independence while helping to accelerate Europe's Industrial Revolution, the extreme profitability of which most certainly cushioned the economic blow of abolition. In sharp contrast, a real end to the fossil fuel age offers no equivalent consolation prizes to the major players in the oil, gas, and coal industries. Solar and wind can make money, sure. But by nature of their decentralization, they will never supply the kind of concentrated super-profits to which the fossil fuel titans have become all too accustomed. In other words, if climate justice carries the day, the economic costs to our elites will be real – not only because of the carbon left in the ground but also because of the regulations, taxes, and social programs needed to make the required transformation. Indeed, these new demands on the ultra rich could effectively bring the era of the footloose Davos oligarch to a close.

The Unfinished Business of Liberation

On one level, the inability of many great social movements to fully realize those parts of their visions that carried the highest price tags can be seen as a cause for inertia or even despair. If they failed in their plans to usher in a more equitable economic system, how can the climate movement hope to succeed?

There is, however, another way of looking at this track record: these economic demands – for basic public services that work, for decent housing, for land redistribution – represent nothing less than the unfinished business of the most powerful liberation movements of the past two centuries, from civil rights to feminism to indigenous sovereignty. The massive global investment required to respond to the climate threat – to adapt humanely and equitably to the heavy weather we have already locked in, and to avert the truly catastrophic warming we can still avoid – is a chance to change all that; and to get it right this time. It could deliver the equitable redistribution of agricultural lands that was supposed to follow independence from colonial rule and dictatorship; it could bring the jobs and homes that Martin Luther King dreamed of; it could bring jobs and clean water to Native communities; it could at last turn on the lights and running water in every South African township. Such is the promise of a Marshall Plan for the Earth.

The fact that our most heroic social justice movements won on the legal front but suffered big losses on the economic front is precisely why our world is as fundamentally unequal and unfair as it remains. Those losses have left a legacy of continued discrimination, double standards, and entrenched poverty – poverty that deepens with each new crisis. But, at the same time, the economic battles the movements *did* win are the reason we still have a few institutions left – from libraries to mass transit to public hospitals – based on the wild idea that real equality means equal access to the basic services that create a dignified life. Most critically, all these past movements, in one form or another, are still fighting today – for full human rights and equality regardless of ethnicity, gender, or sexual orientation; for real decolonization and reparation; for food security and farmers' rights; against oligarchic rule; and to defend and expand the public sphere.

So climate change does not need some shiny new movement that will magically succeed where others failed. Rather, as the furthest-reaching crisis created by the extractivist worldview, and one that puts humanity on a firm and unyielding deadline, climate change can be the force – the grand push – that will bring together all of these still living movements. A rushing river fed by countless streams, gathering collective force to finally reach the sea. 'The basic confrontation which seemed to be colonialism versus anticolonialism, indeed capitalism versus socialism, is already losing its importance,' Frantz Fanon wrote in his 1961 masterwork, *The Wretched of the Earth*. 'What matters today, the issue which blocks the horizon, is the need for a redistribution of wealth. Humanity will have to address this question, no matter how devastating the consequences may be.'[7] Climate change is our chance to right those festering wrongs at last – the unfinished business of liberation.

Winning will certainly take the convergence of diverse constituencies on a scale previously unknown. Because, although there is no perfect historical analogy for the challenge of climate change, there are certainly lessons to learn from the transformative movements of the past. One such lesson is that when major shifts in the economic balance of power take place, they are invariably the result of extraordinary levels of social mobilization. At those junctures, activism becomes something that is not performed by a small tribe within a culture, whether a vanguard of radicals or a subcategory of slick professionals (though each play their part), but becomes an entirely normal activity throughout society – it's rent payers associations, women's auxiliaries, gardening clubs, neighborhood assemblies, trade unions, professional groups, sports teams, youth leagues, and on and on. During extraordinary historical moments – both world wars, the aftermath of the Great Depression, or the peak of the civil rights era – the usual categories dividing 'activists' and 'regular people' became meaningless because the project of changing society was so deeply woven into the project of life. Activists were, quite simply, everyone.

Which brings us back to where we started: climate change and bad timing. It must always be remembered that the greatest barrier to humanity rising to meet the climate crisis is not that it is too late or that we don't know what to do. There is just enough time, and we are swamped with green tech and green plans.

And yet the reason so many of us still feel hopeless is that we are afraid – with good reason – that our political class is wholly incapable of seizing those tools and implementing those plans, since doing so involves unlearning the core tenets of the stifling free-market ideology that governed every stage of their rise to power.

And it's not just the people we vote into office and then complain about – it's us. For most of us living in postindustrial societies, when we see the crackling black-and-white footage of general strikes in the 1930s, victory gardens in the 1940s, and Freedom Rides in the 1960s, we simply cannot imagine being part of any mobilization of that depth and scale. That kind of thing was fine for them but surely not us – with our eyes glued to smart phones, attention spans scattered by click bait, loyalties split by the burdens of debt and insecurities of contract work. Where would we organize? Who would we trust enough to lead us? Who, moreover, is 'we'?

In other words, we are products of our age and of a dominant ideological project. One that too often has taught us to see ourselves as little more than singular, gratification-seeking units, out to maximize our narrow advantage, while simultaneously severing so many of us from the broader communities whose pooled skills are capable of solving problems big and small. This project also has led our governments to stand by helplessly for more than two decades as the climate crisis morphed from a 'grandchildren' problem to a banging-down-the-door problem.

All of this is why any attempt to rise to the climate challenge will be fruitless unless it is understood as part of a much broader battle of worldviews, a process of rebuilding and reinventing the very idea of the collective, the communal, the commons, the civil, and the civic after so many decades of attack and neglect. Because what is overwhelming about the climate challenge is that it requires breaking so many rules at once – rules written into national laws and trade agreements, as well as powerful unwritten rules that tell us that no government can increase taxes and stay in power, or say no to major investments no matter how damaging, or plan to gradually contract those parts of our economies that endanger us all.

And yet each of those rules emerged out of the same, coherent worldview. If that worldview is delegitimized, then all of the rules within it become much weaker and more vulnerable. This is another lesson from social movement history across the political spectrum: when fundamental change does come, it's generally not in legislative dribs and drabs spread out evenly over decades. Rather it comes in spasms of rapid-fire lawmaking, with one breakthrough after another. The right calls this 'shock therapy'; the left calls it 'populism' because it requires so much popular support and mobilization to occur. (Think of the regulatory architecture that emerged in the New Deal period, or, for that matter, the environmental legislation of the 1960s and 1970s.)

So how do you change a worldview, an unquestioned ideology? Part of it involves choosing the right early policy battles – game-changing ones that don't merely aim to change laws but change patterns of thought. That means that a fight for a minimal carbon tax might do a lot less good than, for instance, forming a grand coalition to demand a guaranteed minimum income. That's not only because a minimum income, as discussed, makes it possible for workers to say no to dirty energy jobs but also because the very process of arguing for a universal social safety net opens up a space for a full-throated debate about values – about what we owe to one another based on our shared humanity, and what it is that we collectively value more than economic growth and corporate profits.

Indeed a great deal of the work of deep social change involves having debates during which new stories can be told to replace the ones that have failed us. Because if we are to have any hope of making the kind of civilizational leap required of this fateful decade, we will need to start believing, once again, that humanity is not hopelessly selfish and greedy – the image ceaselessly sold to us by everything from reality shows to neoclassical economics.

Paradoxically, this may also give us a better under-standing of our personal climate inaction, allowing many of us to view past (and present) failures with compassion, rather than angry judgment. What if part of the reason so many of us have failed to act is not because we are too selfish to care about an abstract or seem-ingly far-off problem – but because we are utterly overwhelmed by how much we do care? And what if we stay silent not out of acquiescence but in part because we lack the collective spaces in which to confront the raw terror of ecocide? The end of the world as we know it, after all, is not something anyone should have to face on their own. As the sociologist Kari Norgaard puts it in *Living in Denial*, a fascinating exploration of the way almost all of us suppress the full reality of the climate crisis, 'Denial can – and I believe should – be under-stood as testament to our human capacity for empathy, compassion, and an underlying sense of moral impera-tive to respond, even as we fail to do so.'[8]

Fundamentally, the task is to articulate not just an alternative set of policy proposals but an alternative worldview to rival the one at the heart of the eco-logical crisis – embedded in interdependence rather than hyper-individualism, reciprocity rather than dominance, and cooperation rather than hierarchy. This is required not only to create a political context to dramatically lower emissions, but also to help us cope with the disasters we can no longer hope to avoid. Because in the hot and stormy future we have already made inevitable through our past emissions, an unshakable belief in the equal rights of all people and a capacity for deep compassion will be the only things standing between civilization and barbarism.

This is another lesson from the transformative movements of the past: all of them understood that the process of shifting cultural values – though somewhat ephemeral and difficult to quantify – was central to their work. And so they dreamed in public, showed humanity a better version of itself, modeled different values in their own behavior, and in the process liberated the political imagination and rapidly altered the sense of what was possible. They were also unafraid of the language of morality – to give the pragmatic, cost-benefit arguments a rest and speak of right and wrong, of love and indignation.

In *The Wealth of Nations*, Adam Smith made a case against slavery that had little to do with morality and everything to do with the bottom line. Work by paid laborers, he argued, 'comes cheaper in the end than that performed by slaves': not only were slave owners responsible for the high costs of the 'wear and tear' of their human property but, he claimed, paid laborers had a greater incentive to work hard.[9] Many abolitionists on both sides of the Atlantic would embrace such pragmatic arguments.

However, as the push to abolish the slave trade (and later, slavery itself) ramped up in Britain in the late eighteenth century, much of the movement put considerably more emphasis on the moral travesties of slavery and the corrosive worldview that made it possible. Writing in 1808, British abolitionist Thomas Clarkson described the battle over the slave trade as 'a contest between those who felt deeply for the happiness and the honour of their fellow-creatures, and those who, through vicious custom and the impulse of avarice, had trampled under-foot the sacred rights of their nature, and had even attempted to efface all title to the divine image from their minds.'[10]

The rhetoric and arguments of American abolitionists could be even starker and more uncompromising. In an 1853 speech, the famed abolitionist orator Wendell Phillips insisted on the right to denounce those who in the harshest terms defended slavery. 'Prove to me now that harsh rebuke, indignant denunciation, scathing sarcasm, and pitiless ridicule are wholly and always unjustifiable; else we dare not, in so desperate a case, throw away any weapon which ever broke up the crust of an ignorant prejudice, roused a slumbering conscience, shamed a proud sinner, or changed, in any way, the conduct of a human being. Our aim is to alter public opinion.' And indispensable to that goal were the voices of freed slaves themselves, people like Frederick Douglass, who, in his writing and oratory, challenged the very foundations of American patriotism with questions like 'What, to the American slave, is your 4th of July?'[11]

This kind of fiery, highly polarizing rhetoric was typical of a battle with so much at stake. As the historian David Brion Davis writes, abolitionists understood that their role was not merely to ban an abhorrent practice but to try to change the deeply entrenched values that had made slavery acceptable in the first place. 'The abolition of New World slavery depended in large measure on a major transformation in moral perception – on the emergence of writers, speakers, and reformers, beginning in the mid-eighteenth century, who were willing to condemn an institution that had been sanctioned for thousands of years and who also strove to make human society something more than an endless contest of greed and power.'[12]

This same understanding about the need to assert the intrinsic value of life is at the heart of all major progressive victories, from universal suffrage to universal health care. Though these movements all contained economic arguments as part of building their case for justice, they did not win by putting a monetary value on granting equal rights and freedoms. They won by asserting that those rights and freedoms were *too* valuable to be measured and were inherent to each of us. Similarly, there are plenty of solid economic arguments for moving beyond fossil fuels, as more and more patient investors are realizing. And that's worth pointing out. But we will not win the battle for a stable climate by trying to beat the bean counters at their own game – arguing, for instance, that it is more cost-effective to invest in emission reduction now than disaster response later. We will win by asserting that such calculations are morally monstrous, since they imply that there is an acceptable price for allowing entire countries to disappear, for leaving untold millions to die on parched land, for depriving today's children of their right to live in a world teeming with the wonders and beauties of creation.

The climate movement has yet to find its full moral voice on the world stage, but it is most certainly clearing its throat – beginning to put the very real thefts and torments that ineluctably flow from the decision to mock international climate commitments alongside history's most damned crimes. Some of the voices of moral clarity are coming from the very young, who are calling on the streets and increasingly in the courts for intergenerational justice. Some are coming from great social justice movements of the past, like Nobel laureate Desmond Tutu, former archbishop of Cape Town, who has joined the fossil fuel divestment movement with enthusiasm, declaring that 'to serve as custodians of creation is not an empty title; it requires that we act, and with all the urgency this dire situation demands.'[13] Most of all, those clarion voices are coming from the front lines of the climate crisis, from those lives most directly impacted by both high-risk fossil fuel extraction and early climate destabilization.

Suddenly, Everyone

The 2010s were filled with moments when societies suddenly decided they had had enough, defying all experts and forecasters – from the Arab Spring (tragedies, betrayals, and all), to Europe's 'squares movement' that saw city centers taken over by demonstrators for months, to Occupy Wall Street, to the student movements of Chile and Quebec. The Mexican journalist Luis Hernández Navarro describes those rare political moments that seem to melt cynicism on contact as the 'effervescence of rebellion.'[14]

What is most striking about these upwellings, when societies become consumed with the demand for transformational change, is that they so often come as a surprise – most of all to the movements' own organizers. I've heard the story many times: 'One day it was just me and my friends dreaming up impossible schemes, the next day the entire country seemed to be out in the plaza alongside us.' And the real surprise, for all involved, is that we are so much more than we have been told we are – that we long for more and in that longing have more company than we ever imagined.

No one knows when the next such effervescent moment will open, or whether it will be precipitated by an economic crisis, another natural disaster, or some kind of political scandal. We do know that a warming world will, sadly, provide no shortage of potential sparks. Sivan Kartha, senior scientist at the Stockholm Environment Institute, puts it like this: 'What's politically realistic today may have very little to do with what's politically realistic after another few Hurricane Katrinas and another few Superstorm Sandys and another few Typhoon Bophas hit us.'[15] It's true: the world tends to look a little different when the objects we have worked our whole lives to accumulate are suddenly floating down the street, or smashed to pieces, turned to garbage.

The world also doesn't look much like it did in the late 1980s. Climate change, as we have seen, landed on the public agenda at the peak of free market, end-of-history triumphalism, which was very bad timing indeed. Its do-or-die moment, however, comes to us at a very different historical juncture. Many of the barriers that paralyzed a serious response to the crisis are today significantly eroded. Free market ideology has been discredited by decades of deepening inequality and corruption, stripping it of much of its persuasive power (if not yet its political and economic power). And the various forms of magical thinking that have diverted precious energy – from blind faith in technological miracles to the worship of benevolent billionaires – are also fast losing their grip. It is slowly dawning on a great many of us that no one is going to step in and fix this crisis; that if change is to take place it will only be because leadership bubbled up from below.

We are also significantly less isolated than many of us were even a decade ago: the new structures built in the rubble of neoliberalism – everything from social media to worker co-ops to farmer's markets to neighborhood sharing banks – have helped us to find community despite the fragmentation of postmodern life. Indeed, thanks in particular to social media, a great many of us are continually engaged in a cacophonous global conversation that, however maddening it is at times, is unprecedented in its reach and power.

Given these factors, there is little doubt that another crisis will see us in the streets and squares once again, taking us all by surprise. The real question is what progressive forces will make of that moment, the power and confidence with which it will be seized. Because these moments when the impossible seems suddenly possible are excruciatingly rare and precious. That means more must be made of them. The next time one arises, it must be harnessed not only to denounce the world as it is, and build fleeting pockets of liberated space. It must be the catalyst to actually build the world that will keep us all safe. The stakes are simply too high, and time too short, to settle for anything less.

A year ago, I was having dinner with some newfound friends in Athens. I asked them for ideas about what questions I should put to Alexis Tsipras, the young leader of Greece's official opposition party and one of the few sources of hope in a Europe ravaged by austerity.

Someone suggested, 'Ask him: History knocked on your door, did you answer?'

That's a good question, for all of us.

Notes

Chapter 1: Hot Money

1. Oceanographer Roger Revelle, who led the team that wrote on atmospheric CO_2 in the report for President Johnson, had used similar language describing carbon emissions as a 'geophysical experiment' as early as 1957, in a landmark climate science paper co-authored with chemist Hans Suess: Roger Revelle and Hans E. Suess, 'Carbon Dioxide Exchange Between Atmosphere and Ocean and the Question of an Increase of Atmospheric CO_2 during the Past Decades,' *Tellus* 9 (1957): 19–20. For in-depth histories of climate science and politics, see: Spencer Weart, *The Discovery of Global Warming* (Cambridge, MA: Harvard University Press, 2008); Joshua P. Howe, *Behind the Curve: Science and the Politics of Global Warming* (Seattle: University of Washington Press, 2014). HISTORY: Weart, *The Discovery of Global Warming*, 1–37; JOHNSON REPORT: Roger Revelle et al., 'Atmospheric Carbon Dioxide,' in *Restoring the Quality of Our Environment, Report of the Environmental*

Pollution Panel, President's Science Advisory Committee, The White House, November 1965, Appendix Y4, pp. 126–27.

2. 'Statement of Dr James Hansen, Director, NASA Goddard Institute for Space Studies,' presented to United States Senate, June 23, 1988; Philip Shabecoff, 'Global Warming Has Begun, Expert Tells Senate,' *New York Times*, June 24, 1988; Weart, *The Discovery of Global Warming*, 150–51.

3. Thomas Sancton, 'Planet of the Year: What on EARTH Are We Doing?,' *Time*, January 2, 1989.

4. *Ibid.*

5. President R. Venkataraman, 'Towards a Greener World,' speech at WWF-India, New Delhi, November 3, 1989, in *Selected Speeches, Volume I: July 1987–December 1989* (New Delhi: Government of India, 1991), 612.

6. Daniel Indiviglio, 'How Americans' Love Affair with Debt Has Grown,' *The Atlantic*, September 26, 2010.

7. One bold proposal imagines future restrictions on trade in all goods produced with fossil fuels, arguing that once the green transition is underway and industries have begun to decarbonize, such measures could be introduced and ramped up gradually: Tilman Santarius, 'Climate and Trade: Why Climate Change Calls for Fundamental Reforms in World Trade Policies,' German NGO Forum on Environment and Development, Heinrich Böll Foundation, pp. 21–23. U.N. CLIMATE AGREEMENT: United Nations Framework Convention on Climate Change, United Nations, 1992, Article 3,

Principle 5; 'PIVOTAL MOMENT': Robyn Eckersley, 'Understanding the Interplay Between the Climate and Trade Regimes,' in *Climate and Trade Policies in a Post-2012 World*, United Nations Environment Programme, p. 17.

8. Martin Khor, 'Disappointment and Hope as Rio Summit Ends,' in *Earth Summit Briefings* (Penang: Third World Network, 1992), p. 83.

9. Steven Shrybman, 'Trade, Agriculture, and Climate Change: How Agricultural Trade Policies Fuel Climate Change,' Institute for Agriculture and Trade Policy, November 2000, p. 1.

10. Sonja J. Vermeulen, Bruce M. Campbell, and John S.I. Ingram, 'Climate Change and Food Systems,' *Annual Review of Environment* 37 (2012): 195; personal email communication with Steven Shrybman, April 23, 2014.

11. 'Secret Trans-Pacific Partnership Agreement (TPP) – Environment Consolidated Text,' WikiLeaks, January 15, 2014, https://wikileaks.org; 'Summary of U.S. Counterproposal to Consolidated Text of the Environment Chapter,' released by RedGE, February 17, 2014, http://www.redge.org.pe.

12. Traffic refers to containerized port traffic, measured by twenty-foot equivalent units (TEUs). From 1994 to 2013 global containerized port traffic increased from 128,320,326 to an estimated 627,930,960 TEUs, an increase of 389.4 percent: United Nations Conference on Trade and Development, 'Review of Maritime Transport,' various years, available at http://unctad.

org. For years 2012 and 2013, port traffic was projected based on industry estimates from Drewry: 'Container Market Annual Review and Forecast 2013/14,' Drewry, October 2013. NOT ATTRIBUTED: 'Emissions from Fuel Used for International Aviation and Maritime Transport (International Bunker Fuels),' United Nations Framework Convention on Climate Change, http://unfccc.int; SHIPPING EMMISSIONS: Øyvind Buhaug et al., 'Second IMO GHG Study 2009,' International Maritime Organization, 2009, p. 1.

13. 'European Union CO_2 Emissions: Different Accounting Perspectives,' European Environmental Agency Technical Report No. 20/2013, 2013, pp. 7–8.

14. Glen P. Peters et al., 'Growth in Emission Transfers via International Trade from 1990 to 2008,' *Proceedings of the National Academy of Sciences* 108 (2011): 8903-4.

15. Corrine Le Quéré et al., 'Global Budget 2013,' *Earth System Science Data* 6 (2014): 252; Corrine Le Quéré et al., 'Trends in the Sources and Sinks of Carbon Dioxide,' *Nature Geoscience* 2 (2009): 831; Ross Garnaut et al., 'Emissions in the Platinum Age: The Implications of Rapid Development for Climate-Change Mitigation,' *Oxford Review of Economic Policy* 24 (2008): 392; Glen P. Peters et al., 'Rapid Growth in CO_2 Emissions After the 2008–2009 Global Financial Crisis,' *Nature Climate Change* 2 (2012): 2; 'Technical Summary,' in O. Edenhofer et al., ed., *Climate Change 2014: Mitigation of Climate Change, Contribution of Working Group III to the Fifth Assessment Report of the Intergovernmental Panel on*

Climate Change (Cambridge: Cambridge University Press), 15.

16. Andreas Malm, 'China as Chimney of the World: The Fossil Capital Hypothesis,' *Organization & Environment* 25 (2012): 146, 165; Yan Yunfeng and Yang Laike, 'China's Foreign Trade and Climate Change: A Case Study of CO_2 Emissions,' *Energy Policy* 38 (2010): 351; Ming Xu et al., 'CO_2 Emissions Embodied in China's Exports from 2002 to 2008: A Structural Decomposition Analysis,' *Energy Policy* 39 (2011): 7383.

17. Personal interview with Margrete Strand Rangnes, March 18, 2013.

18. Malm, 'China as Chimney of the World,' 147, 162.

19. Elisabeth Rosenthal, 'Europe Turns Back to Coal, Raising Climate Fears,' *New York Times*, April 23, 2008; Personal email communication with IEA Clean Coal Centre, March 19, 2014.

20. Jonathan Watts, 'Foxconn offers pay rises and suicide nets as fears grow over wave of deaths,' *Guardian*, May 28, 2010; Shahnaz Parveen, 'Rana Plaza factory collapse survivors struggle one year on,' BBC News, April 23, 2014.

21. Mark Dowie, *Losing Ground: American Environmentalism at the Close of the Twentieth Century* (Cambridge, MA: MIT Press, 1996), 185-86; Keith Schneider, 'Environment Groups Are Split on Support for Free-Trade Pact,' *New York Times*, September 16, 1993.

22. Dowie, *Losing Ground*, 186–87; Gilbert A. Lewthwaite, 'Gephardt Declares Against NAFTA; Democrat Cites Threat to U.S. Jobs,' *Baltimore Sun*, September 22, 1993;

John Dillin, 'NAFTA Opponents Dig In Despite Lobbying Effort,' *Christian Science Monitor*, October 12, 1993; Mark Dowie, 'The Selling (Out) of the Greens; Friends of Earth – or Bill?' *The Nation*, April 18, 1994.

23. Bill Clinton, 'Remarks on the Signing of NAFTA (December 8, 1993),' Miller Center, University of Virginia.

24. Stan Cox, 'Does It Really Matter Whether Your Food Was Produced Locally?' *Alternet*, February 19, 2010.

25. Solomon interview, August 27, 2013.

26. Kevin Anderson, 'Climate Change: Going Beyond Dangerous – Brutal Numbers and Tenuous Hope,' *Development Dialogue* no. 61, September 2012, pp. 16-40.

27. The '8 to 10 percent' range relies on interviews with Anderson and Bows-Larkin as well as their published work. For the underlying emissions scenarios, refer to pathways C+1, C+3, C+5, and B6 3 in: Kevin Anderson and Alice Bows, 'Beyond "Dangerous" Climate Change: Emission Scenarios for a New World,' *Philosophical Transactions of the Royal Society A* 369 (2011): 35. See also: Kevin Anderson, 'EU 2030 Decarbonisation Targets and UK Carbon Budgets: Why So Little Science?' KevinAnderson.info, June 14, 2013, http://kevinanderson.info. DE BOER: Alex Morales, 'Kyoto Veterans Say Global Warming Goal Slipping Away,' Bloomberg, November 4, 2013.

28. Stern, *The Economics of Climate Change*, 231–32.

29. *Ibid.*, 231; Global Carbon Project emissions data, 2013 Budget v2.4 (July 2014), available at http://cdiac.ornl.

gov; Carbon Dioxide Information Analysis Center emissions data, available at http://cdiac.ornl.gov.

30. Kevin Anderson and Alice Bows, 'A 2°C Target? Get Real, Because 4°C Is on Its Way,' *Parliamentary Brief* 13 (2010): 19; FOOTNOTE: Anderson and Bows, 'Beyond "Dangerous" Climate Change,' 35; Kevin Anderson, 'Avoiding Dangerous Climate Change Demands De-growth Strategies from Wealthier Nations,' KevinAnderson.info, November 25, 2013, http://kevin-anderson.info.

31. Anderson and Bows-Larkin have based their analysis on the commitment made by governments at the 2009 U.N. climate summit in Copenhagen that emission cuts should be done on the basis 'of equity' (meaning rich countries must lead so that poor countries have room to develop). Some argue that rich countries don't have to cut quite so much. Even if that were true, however, the basic global picture still suggests that the necessary reductions are incompatible with economic growth as we have known it. As Tim Jackson shows in *Prosperity Without Growth*, global annual emission cuts of as little as 4.9 percent cannot be achieved simply with green tech and greater efficiencies. Indeed he writes that to meet that target, with the world population and income per capita continuing to grow at current rates, the carbon intensity of economic activity would need to go down 'almost ten times faster than it is doing right now.' And by 2050, we would need to be twenty-one times more efficient than we are

today. So, even if Anderson and Bows-Larkin have vastly overshot, they are still right on their fundamental point: we need to change our current model of growth. See: Tim Jackson, *Prosperity Without Growth: Economics for a Finite Planet* (London: Earthscan, 2009): 80, 86.

32. Anderson and Bows, 'A New Paradigm for Climate Change,' 640.

33. Kevin Anderson, 'Romm Misunderstands Klein's and My View of Climate Change and Economic Growth,' KevinAnderson.info, September 24, 2013.

34. Clive Hamilton, 'What History Can Teach Us About Climate Change Denial,' in *Engaging with Climate Change: Psychoanalytic and Interdisciplinary Perspectives*, ed. Sally Weintrobe (East Sussex: Routledge, 2013), 18.

35. For the foundational scenario modeling work on a 'Great Transition' to global sustainability, led by researchers at the Tellus Institute and the Stockholm Environment Institute, see: Paul Raskin et al., 'Great Transition: The Promise and Lure of the Times Ahead,' Report of the *Global Scenario Group*, Stockholm Environment Institute and Tellus Institute, 2002. This research has continued as part of Tellus' Great Transition Initiative, available at: 'Great Transition Initiative: Toward a Transformative Vision and Praxis,' Tellus Institute, http://www.greattransition.org. For parallel work at the U.K.'s New Economics Foundation, see: Stephen Spratt, Andrew Simms, Eva Neitzert, and Josh Ryan-Collins, 'The Great Transition,' The New Economics Foundation, June 2010.

36. Bows interview, January 14, 2013.

37. Rebecca Willis and Nick Eyre, 'Demanding Less: Why We Need a New Politics of Energy,' Green Alliance, October 2011, pp. 9, 26.

38. FOOTNOTE: 'EP Opens Option for a Common Charger for Mobile Phones,' European Commission, press release, March 13, 2014; Adam Minter, *Junkyard Planet* (New York: Bloomsbury, 2013), 6–7, 67, 70.

39. This quote has been clarified slightly at Anderson's request. Paul Moseley and Patrick Byrne, 'Climate Expert Targets the Affluent,' BBC, November 13, 2009.

40. Phaedra Ellis-Lamkins, 'How Climate Change Affects People of Color,' *The Root*, March 3, 2013.

41. Tim Jackson, 'Let's Be Less Productive,' *New York Times*, May 26, 2012.

42. John Stutz, 'Climate Change, Development and the Three-Day Week,' Tellus Institute, January 2, 2008, pp. 4-5. See also: Juliet B. Schor, *Plenitude: The New Economics of True Wealth* (New York: Penguin Press, 2010); Kyle W. Knight, Eugene A. Rosa, and Juliet B. Schor, 'Could Working Less Reduce Pressures on the Environment? A Cross-National Panel Analysis of OECD Countries, 1970–2007,' *Global Environmental Change* 23 (2013): 691-700.

43. Alyssa Battistoni, 'Alive in the Sunshine,' *Jacobin* 13 (2014): 25.

Chapter 2: Public and Paid For

1. Personal interview with Nastaran Mohit, November 10, 2012.

2. Steve Kastenbaum, 'Relief from Hurricane Sandy Slow for Some,' CNN, November 3, 2012.

3. Johnathan Mahler, 'How the Coastline Became a Place to Put the Poor,' *New York Times*, December 3, 2012; personal interview with Aria Doe, executive director, Action Center for Education and Community Development, February 3, 2013.

4. Sarah Maslin Nir, 'Down to One Hospital, Rockaway Braces for Summer Crowds,' *New York Times*, May 20, 2012; personal email communication with Nastaran Mohit, March 28, 2014; Mohit interview, November 10, 2012.

5. *Ibid.*; FOOTNOTE: Greg B. Smith, 'NYCHA Under Fire for Abandoning Tenants in Hurricane Sandy Aftermath,' New York *Daily News*, November 19, 2012.

6. Mohit interview, November 10, 2012.

7. *Ibid.*

8. Andrew P. Wilper et. al., 'Health Insurance and Mortality in U.S. Adults,' *American Journal of Public Health* 99 (2009): 2289–95; Mohit, November 10, 2012.

9. Aria Doe interview, February 3, 2013.

10. John Aglionby, Mark Odell, and James Pickford, 'Tens of Thousands Without Power After Storm Hits Western Britain,' *Financial Times*, February 13, 2014; Tom

Bawden, 'St Jude's Day Storm: Four Dead After 99mph Winds and Night of Destruction – But at Least We Saw It Coming,' *The Independent* (London), October 29, 2013.

11. Alex Marshall, 'Environment Agency Cuts: Surviving the Surgeon's Knife,' *The ENDS Report*, January 3, 2014; Damian Carrington, 'Massive Cuts Risk England's Ability to Deal with Floods, MPs Say,' *Guardian*, January 7, 2014; Damian Carrington, 'Hundreds of UK Flood Defence Schemes Unbuilt Due to Budget Cuts,' *Guardian*, July 13, 2012.

12. Dave Prentis, 'Environment Agency Workers Are Unsung Heroes,' UNISON, January 6, 2014.

13. EM-DAT, International Disaster Database, Centre for Research on the Epidemiology of Disasters (advanced searches), http://www.emdat.be/database; personal email communication with Michael Mann, March 27, 2014.

14. 'Billion-Dollar Weather/Climate Disasters,' National Climatic Data Center, http://www.ncdc.noaa.gov; 'Review of Natural Catastrophes in 2011: Earthquakes Result in Record Loss Year,' Munich RE, press release, January 4, 2012.

15. Personal interview with Amy Bach, September 18, 2012.

16. 'Climate Change: Impacts, Vulnerabilities and Adaptation in Developing Countries,' UNFCCC, 2007, pp. 18–26, 29–38; 'Agriculture Needs to Become "Climate-Smart,"' Food and Agriculture Organization of the UN, October 28, 2010.

17. 'World Economic and Social Survey 2011: The Great Green Technological Transformation,' United Nations Department of Economic and Social Affairs, 2011, pp. xxii, 174.

18. The oil and gas sector was either the most represented or tied for the most represented sector in the top 20 of Fortune's Global 500 rankings for 2012 and 2013: 'Fortune Global 500,' CNN Money, 2013, http://money.cnn.com; 'Fortune Global 500,' CNN Money, 2012, http://money.cnn.com. BLOCKED PROGRESS: James Hoggan with Richard Littlemore, *Climate Cover-Up: The Crusade to Deny Global Warming* (Vancouver: Greystone Books, 2009); $900 BILLION: Daniel J. Weiss, 'Big Oil's Lust for Tax Loopholes,' Center for American Progress, January 31, 2011; 2011 EARNINGS: '2011 Summary Annual Report,' ExxonMobil, p. 4; 2012 EARNINGS: '2012 Summary Annual Report,' ExxonMobil, p. 4; 'Exxon's 2012 Profit of $44.9B Misses Record,' Associated Press, February 1, 2013.

19. BP, for instance, pledged $8 billion for alternative energy in 2005. Saaed Shah, 'BP Looks "Beyond Petroleum" with $8bn Renewables Spend,' *The Independent* (London), November 29, 2005; BEYOND PETROLEUM: Terry Macalister and Eleanor Cross, 'BP Rebrands on a Global Scale,' *Guardian*, July 24, 2000; HELIOS MARK: 'BP Amoco Unveils New Global Brand to Drive Growth,' press release, July 24, 2000; BROWNE: Terry Macalister and Eleanor Cross, 'BP Rebrands on a Global Scale,' *Guardian*, July 24, 2000;

CHEVRON: 'We Agree: Oil Companies Should Support Renewable Energy' (video), Chevron, YouTube, 2010; 2009 STUDY: Daniel J. Weiss and Alexandra Kougentakis, 'Big Oil Misers,' Center for American Progress, March 31, 2009; EXECUTIVE PAY: James Osborne, 'Exxon Mobil CEO Rex Tillerson Gets 15 Percent Raise to $40.3 Million,' *Dallas Morning News*, April 12, 2013.

20. Antonia Juhasz, 'Big Oil's Lies About Alternative Energy,' *Rolling Stone*, June 25, 2013; Ben Elgin, 'Chevron Dims the Lights on Green Power,' *Bloomberg Businessweek*, May 29, 2014; Ben Elgin, 'Chevron Backpedals Again on Renewable Energy,' *Bloomberg Businessweek*, June 9, 2014.

21. Brett Martel, 'Jury Finds Big Tobacco Must Pay $590 Million for Stop-Smoking Programs,' Associated Press, May 21, 2004; Bruce Alpert, 'U.S. Supreme Court Keeps Louisiana's $240 Million Smoking Cessation Program Intact,' *Times-Picayune*, June 27, 2011; Sheila McNulty and Ed Crooks, 'BP Oil Spill Pay-outs Hit $5bn Mark,' *Financial Times*, August 23, 2011; Lee Howell, 'Global Risks 2013,' World Economic Forum, 2013, p. 19.

22. Marc Lee, 'Building a Fair and Effective Carbon Tax to Meet BC's Greenhouse Gas Targets,' Canadian Centre for Policy Alternatives, August 2012.

23. U.S. Department of Defense emissions were calculated using the federal Greenhouse Gas Inventory for fiscal year 2011 (total Scope 1 emissions, excluding biogenic). 'Fiscal Year 2011 Greenhouse Gas Inventory,'

U.S. Department of Energy, Office of Energy Efficiency and Renewable Energy, June 14, 2013, http://energy.gov; 'Greenhouse Gas 100 Polluters Index,' Political Economy Research Institute, University of Massachusetts Amherst, June 2013, http://www.peri.umass.edu.

24. Borgar Aamaas, Jens Borken-Kleefeld, and Glen P. Peters, 'The Climate Impact of Travel Behavior: A German Case Study with Illustrative Mitigation Options,' *Environmental Science & Policy* 33 (2013): 273, 276.

25. Thomas Piketty, *Capital in the Twenty-First Century*, trans. Arthur Goldhammer (Cambridge, MA: Harvard University Press, 2014); Gar Lipow, *Solving the Climate Crisis through Social Change: Public Investment in Social Prosperity to Cool a Fevered Planet* (Santa Barbara: Praeger, 2012), 56; Stephen W. Pacala, 'Equitable Solutions to Greenhouse Warming: On the Distribution of Wealth, Emissions and Responsibility Within and Between Nations,' presentation to International Institute for Applied Systems Analysis, November 2007, p. 3.

26. 'Innovative Financing at a Global and European Level,' European Parliament, resolution, March 8, 2011, http://www.europarl.europa.eu.

27. 'Revealed: Global Super-Rich Has at Least $21 Trillion Hidden in Secret Tax Havens,' Tax Justice Network, press release, July 22, 2012.

28. 'World Economic and Social Survey 2012: In Search of New Development Finance,' United Nations Department of Economic and Social Affairs, 2012, p. 44.

29. Sam Perlo-Freeman, et. al., 'Trends in World Military Expenditure, 2012,' Stockholm International Peace Research Institute, April 2013, http://sipri.org.

30. 'Mobilizing Climate Finance: A Paper Prepared at the Request of G20 Finance Ministers,' World Bank Group, October 6, 2011, p.15, http://www.imf.org.

31. 'Governments Should Phase Out Fossil Fuel Subsidies or Risk Lower Economic Growth, Delayed Investment in Clean Energy and Unnecessary Climate Change Pollution,' Oil Change International and Natural Resources Defense Council, June 2012, p. 2.

32. For a more in-depth, U.S.-focused discussion of raising climate funds from these kinds of sources, see Lipow, *Solving the Climate Crisis through Social Change*, 55-61.

33. For much more on rationing, climate change, and environmental and economic justice, see: Stan Cox, *Any Way You Slice It: The Past, Present, and Future of Rationing* (New York: The New Press, 2013). 16 PERCENT: Ina Zweiniger-Bargielowska, *Austerity in Britain: Rationing, Controls, and Consumption, 1939–1955* (Oxford: Oxford University Press, 2000), 55, 58.

34. Nicholas Timmins, 'When Britain Demanded Fair Shares for All,' *The Independent* (London), July 27, 1995; Martin J. Manning and Clarence R. Wyatt, *Encyclopedia of Media and Propaganda in Wartime America,* Vol. 1 (Santa Barbara, CA: ABC-CLIO: 2011), 533; Terrence H. Witkowski, 'The American Consumer Home Front During World War II,' *Advances in Consumer Research* 25 (1998).

35. *Rationing, How and Why?* pamphlet, Office of Price Administration, 1942, p. 3.

36. Donald Thomas, *The Enemy Within: Hucksters, Racketeers, Deserters and Civilians During the Second World War* (New York: New York University Press, 2003), 29; Hugh Rockoff, *Drastic Measures: A History of Wage and Price Controls in the United States* (Cambridge: Cambridge University Press, 1984), 166–67.

37. Jimmy Carter, 'Crisis of Confidence' speech (transcript), *American Experience*, PBS.

38. 'The Pursuit of Progress' (video), *Richard Heffner's Open Mind*, PBS, February 10, 1991.

39. Eleanor Taylor, 'British Social Attitudes 28,' Chapter 6, Environment, NatCen Social Research, p. 104.

40. Will Dahlgreen, 'Broad Support for 50P Tax,' YouGov, January 28, 2014; 'Nine in Ten Canadians Support Taxing the Rich "More" (88%) and a Potential "Millionaire's Tax" (89%),' Ipsos, May 30, 2013; Anthony Leiserowitz et al., 'Public Support for Climate and Energy Policies in November 2013,' Yale Project on Climate Change Communication and George Mason University Center for Climate Change Communication, November 2013; 'Voter Attitudes Toward Pricing Carbon and a Clean Energy Refund' (memo), Public Opinion Strategies, April 21, 2010.

41. 'Americans Support Limits on CO_2,' Yale Project on Climate Change Communication, April 2014.

Chapter 3: The Leap Years

1. Martin Luther King Jr, *Where Do We Go from Here: Chaos or Community?* (Boston: Beacon, [1967] 2010), 5–6.
2. Johannes G. Hoogeveen and Berk Özler, 'Not Separate, Not Equal: Poverty and Inequality in Post-Apartheid South Africa,' Working Paper No. 739, William Davidson Institute, University of Michigan Business School, January 2005.
3. For work exploring the multi-faceted parallels between climate change, slavery, and abolitionism more broadly, see: Jean-François Mouhot, 'Past Connections and Present Similarities in Slave Ownership and Fossil Fuel Usage,' *Climatic Change* 105 (2011): 329–355; Jean-François Mouhot, *Des Esclaves Énergétiques: Réflexions sur le Changement Climatique* (Seyssel: Champ Vallon, 2011); Andrew Nikiforuk, *The Energy of Slaves* (Vancouver: Greystone Books, 2012); HAYES: Christopher Hayes, 'The New Abolitionism,' *The Nation*, April 22, 2014.
4. Greg Grandin, 'The Bleached Bones of the Dead: What the Modern World Owes Slavery (It's More Than Back Wages),' TomDispatch, February 23, 2014; Adam Hochschild, *Bury the Chains: Prophets and Rebels in the Fight to Free an Empire's Slaves* (New York: Houghton Mifflin, 2006), 13–14, 54–55.
5. Christopher Hayes, 'The New Abolitionism,' *The Nation*, April 22, 2014; FOOTNOTE: Seth Rockman

and Sven Beckert, eds., *Slavery's Capitalism: A New History of American Economic Development* (Philadelphia: University of Pennsylvania Press, Sven Beckert and Seth Rockman, 'Partners in Iniquity,' *New York Times*, April 2, 2011; Julia Ott, 'Slaves: The Capital That Made Capitalism,' Public Seminar, April 9, 2014; Edward E. Baptist and Louis Hyman, 'American Finance Grew on the Back of Slaves,' *Chicago Sun-Times*, March 7, 2014; Katie Johnston, 'The Messy Link Between Slave Owners and Modern Management,' *Forbes*, January 16, 2013.

6. Lauren Dubois, *Haiti: The Aftershocks of History* (New York: Metropolitan Books, 2012), 97–100.

7. Frantz Fanon, *The Wretched of the Earth* (New York: Grove, 2004), 55.

8. Kari Marie Norgaard, *Living in Denial: Climate Change, Emotions, and Everyday Life* (Cambridge, MA: MIT Press, 2011), 61.

9. Adam Smith, *The Wealth of Nations*, Books I–III, ed. Andrew Skinner (London: Penguin, 1999), 183–84, 488–89.

10. Seymour Drescher, *The Mighty Experiment: Free Labor Versus Slavery in British Emancipation* (Oxford: Oxford University Press, 2002), 34–35, 233; Thomas Clarkson, *The History of the Rise, Progress, and Accomplishment of the Abolition of the African Slave-Trade, by the British Parliament*, Vol. 2 (London: Longman, Hurst, Rees, and Orme, 1808), 580–81.

11. Wendell Phillips, 'Philosophy of the Abolition Movement: Speech Before the Massachusetts Antislavery Society (1853),' in *Speeches, Lectures, and Letters* (Boston: James Redpath, 1863), 109–10; Frederick Douglass, 'The Meaning of July Fourth for the Negro,' speech at Rochester, New York, July 5, 1852, in *Frederick Douglass: Selected Speeches and Writings*, ed. Philip S. Foner and Yuval Taylor (Chicago: Chicago Review Press, 2000), 196.
12. David Brion Davis, *Inhuman Bondage: The Rise and Fall of Slavery in the New World* (New York: Oxford University Press, 2006), 1.
13. Desmond Tutu, 'We Need an Apartheid-Style Boycott to Save the Planet,' *Guardian*, April 10, 2014.
14. Luis Hernández Navarro, 'Repression and Resistance in Oaxaca,' *CounterPunch*, November 21, 2006.
15. Personal interview with Sivan Kartha, January 11, 2013.